EASTERN NIGERIA

Eastern Nigeria

A Geographical Review

Barry Floyd

FREDERICK A. PRAEGER, *Publishers*

New York · Washington

BOOKS THAT MATTER

Published in the United States of America in 1969 by
FREDERICK A. PRAEGER, INC., *Publishers*
111 Fourth Avenue, New York, N.Y. 10003

© 1969, in London, England, by Barry Floyd

Library of Congress Catalog Card Number: 71-78092

Printed in Great Britain

Dedicated to the foundation students in Geography at the University of Nigeria, in memory of Nsukka days and Cross River nights

Contents

List of Plates

The author and publishers wish to thank the following, who have given permission for the photographs to be reproduced:

Crisp Pictures: plate 44; Ministry of Agriculture, Enugu: plates 20, 21, 22, 24, 25, 26 and 28; Ministry of Information, Enugu: plates 11, 39, 41, 42, 43, 45, 46, 51 and 52; B. A. Molski: plates 1, 2, 3, 4, 5, 6, 7, 8, 9, 10, 12, 13, 14, 15, 18, 19, 23, 27, 29, 30, 31, 32, 33, 47, 48, 50 and 53; Nigerian Cement Company: plate 40; Shell–BP: plates 34, 35, 36, 37, 38 and 49.

Photographs 16 and 17 were taken by the author.

Preface

THIS study is a systematic geographical appraisal of Eastern Nigeria prior to the outbreak of the Nigerian Civil War in mid-1967. It is based upon four years of teaching and research at the University of Nigeria, Nsukka, from 1962 to 1966, at a time when the Eastern Region (within the framework of the Federal Republic of Nigeria) was making firm strides towards its goal of elevated social conditions and economic advancement for its peoples: Ibo, Ibibio-Efik, Ibibio and others.

Yet no attempt has been made in this book to analyse the bitter conflict between secessionist Eastern Nigeria and the Federal Military Government, or to illustrate geographically the centrifugal forces which have torn the country asunder. There are (and will be) enough studies appearing on this complex issue from the pens of political scientists, historians, sociologists, psychologists and others to satisfy our quest for fuller understanding of the causes of the war.

Rather does this book set out to provide an objective, scientific review of the spatial distribution of both human and physical resources in Eastern Nigeria under 'normal' or peace-time circumstances. It portrays the realities of Eastern Region geography before the Civil War, and in this sense may perhaps be viewed as a case-study in historical geography. At the same time the study sets the stage for the formidable task of reconstruction which clearly lies ahead in Eastern Nigeria, after peace is restored; for both the region's and the nation's wounds must somehow be healed.

We may assume that first efforts will be aimed at re-creating and re-establishing the former patterns of man–land relationships in Eastern Nigeria, whether in the realms of agriculture, commerce, industry, or rural and urban settlement. To this end, the geographical review of Eastern Nigeria provides a 'stock-take' of those physical and socio-economic facts of the area which can be used as a basis for the rehabilitation and resettlement of Eastern Nigerians.

Nevertheless, it will not be possible fully to re-establish the *status quo*, even if this were an entirely desirable objective. One of the few benefits which stem from warfare, following the restoration of peace, is an opportunity to take a new and hard look at the old, established ways of doing things: of producing food, of owning the land, of travelling and trading, of manufacturing, of planning for future growth.

The new architects of the Eastern Nigerian edifice will thus have an opportunity of reviewing pre-war patterns of human geography in the region, and of ameliorating certain of the anomalies revealed by these patterns, as described within this book. They will also have a chance to redeem earlier errors in development planning, minimising in particular certain regrettable consequences of former political decision-making in the location of, for example, new industries, agricultural projects and housing schemes.

But we are still assuming that, with the restoration of peace, the stirred-up complex of physical and social ingredients in the Eastern Nigerian picture will settle back into something like its former arrangement. We should recognise that the complex itself, at least on the human side, has been much altered, particularly by the mass exodus from Northern and Western Nigeria, also the Federal capital of Lagos, of some two million refugees. The return of these 'sons abroad' to the Ibo 'heartland' has placed excessive strain upon the food-producing capabilities of their birth-place villages where – in many areas – over-population and soil degradation had already induced tell-tale signs of deterioration upon the landscape by 1966. Furthermore, the valuable remittances home from the 'sons abroad', with which many family and community development projects were funded in earlier years, have now dried up; also there is insufficient creative work for the talents and experience of the returned Eastern emigrants, many of whom are accustomed to 'big-city' life in Lagos, Ibadan, Benin, Kano and elsewhere in the Federation.

Thus, not only do many of the basic problems of the development of human and land resources in Eastern Nigeria (which have been analysed in this book) remain from the pre-war years; they have been much exacerbated by the present crisis in the political life of the country. One is confident that indigenous and expatriate planners required for the reconstruction period after hostilities cease will find this study a useful benchmark for their deliberations on the future course of the region's life. Ideally, however, a new study of the geography of Eastern Nigeria should be produced a few years hence, after the dust has settled, and the changing patterns of social and economic phenomena have begun to emerge.

So much for the pragmatic and practical applications of this book to planning for the future of the Eastern Provinces. There is still the educational dimension to comment upon, which was in fact the primary focus of the work when it was first conceived. I should like to revert now to the prefatory remarks which were written for *Eastern Nigeria: a Geographical Review* in 1963, at a time and place which appear almost halcyon and idyllic in retrospect.

The silver extent of the broad, meandering Cross River near Obubra in Eastern Nigeria, viewed from the District Officer's hill in the late afternoon sun, provides a charming epitome of a typical African river, yet it reminds one curiously of the Thames Valley in Surrey on a fine summer's day, or of the lower Connecticut River in New England under comparable weather conditions.

Standing on the brink of a great gash in the dark red-earth soils at the Agula-Nanka erosion gullies, in Awka Division, memories are summoned of similar scenes in the former worked-out lands of the Southern Appalachian piedmont or the erosion-scarred farms on the High Veld in South Africa.

Atop a laterite-capped residual hill in the Nsukka-Okigwi Uplands corresponding landscapes are recalled in the Deccan Plateau of Southern India and in the Mashonaland reserves in Rhodesia, even if the materials and agents of landform evolution have differed and the cultural overlay is dissimilar.

Observing the essential activity of men and women traders at the vast Onitsha emporium on the Niger is to be transported to the crowded, covered market in Port-au-Prince, Haiti, or to the teeming multifarious stalls along Chowringee Road in Calcutta.

In the writer's opinion, this is the very essence and stuff of geography as an educational subject: the sensitive observation, description and interpretation of landscapes, land-use patterns and populations, and the attempt to unravel the causes and pursue the consequences of similarities and differences in these earthly characteristics as they appear from place to place within a country, or from region to region around the world.

Re-expressing these aims, it may be said that what a geographer tries to do is to bring the study of man, quite literally, down to earth.

> In so doing, he claims not to be able to offer a self-contained system of explanations for the terrestial distribution of man and his culture, but to provide a vantage point from which to regard these things – a vantage point from which he can examine objectively the 'living tether' by which earth, man and society are irrevocably bound.[1]

[1] G. H. T. Kimble, *The Way of the World* (New York, 1953), p. ix.

The educational purpose of this book is thus to establish a geographical vantage point from which to observe the Eastern Nigerian scene and whence to examine in a systematic fashion those compound ingredients which together make up the geographical 'personality' of the region.

A further observation on the nature of geography is relevant to the academic aims and organisation of the present book. I firmly believe that modern geography should be a man-centred science. Its essential concern should be with the spatial characteristics of human populations on the earth's surface, with the object of understanding why men live where they do and in what ways they earn their living. Location, size and the socio-economic functions of population groups provide the most satisfactory central theme for geography; population patterns are the most appropriate nexus or meeting-place for the several branches of geography as an educational subject.

Population studies offer the touchstone of relevance to our geographic endeavours and ensure a vital connecting theme to our scholarly enquiries. It is for these reasons that the following study departs from the traditional structure of most regional textbooks. Rather than first surveying the physical realities of Eastern Nigeria, attention is directed to the people of the region, particularly to their ethnic diversities and the remarkable disparities in their areal distribution. With these in mind, an account of the physical environment then follows, and an attempt is made to determine the influence of natural factors upon population distribution. The third section of the book covers the principal economic activities of Eastern Nigerians and relates these to the previously-described patterns of human and physical geography. Throughout the work, the connecting theme of the characteristics of population is clearly evident.

The overall purpose of the book then is to provide a general description and appreciation of the facts of Eastern Nigeria's geography, a geographical review of the Eastern Provinces suitable for Nigerian students in the sixth forms of Secondary Schools, in Teachers' Training Colleges, and in Nigerian and other West African Universities. A further potential reader is the expatriate layman and short-term resident in Nigeria who is seeking a straightforward and informative overview of physical and cultural patterns within the Eastern Region. Beyond Africa – in Europe, North America and elsewhere – those with a professional concern or personal interest in African affairs should also find this study helpful as a general reference work.

The opportunity is taken to acknowledge with gratitude the research grants received from Michigan State University and the Economic

Development Institute, University of Nigeria, for pursuing this study. Without them, the field-work which acted as a matrix for the author's knowledge and appreciation of Eastern Nigeria could not have been accomplished. Many individuals have also added to the store of information which was acquired over our four years of residence in Nigeria. To mention each by name is neither feasible nor apposite for the list would be quite lengthy and, furthermore, incomplete. The names of many informants have gone unrecorded for they were the people whom one met in villages and towns, in shops and within the open markets, in farm lands and on forest paths, aboard ferries and in the 'Catering Rest Houses' (now the 'Progress Hotels'). I should like to think that they are all still alive, and none has met with a violent death; but I fear the basis for expressing such a hope is a frail one.

A number of staff members and students of the University of Nigeria, also government officials, were approached for specific information and advice; their contributions have been suitably acknowledged in the appropriate portions of the book. The encouragement of colleagues and students in the Geography Department was particularly appreciated.

My wife was of immeasurable aid in ways too numerous to mention; our family of (then) four young children had occasion to wonder at times whether their father actually existed! A loyal assistant succumbed to malaria and was forced to return to the U.S.A. after just a few months of residence in Nigeria. Her pioneer secretarial efforts were nevertheless greatly appreciated. I should also like to express appreciation for the skilful, sympathetic handling of the draft for this book on the part of Macmillan's Geography Editor, Miss Juliet Williams. She exhibited the utmost professional patience and tact when faced with the task of reducing the exuberant outflow of academic endeavours into a commercially feasible publication. Arthur Banks is to be congratulated on the excellence of the cartographical illustrations. The author's sincere thanks to all participants in the project are hereby recorded.

Kingston, Jamaica   BARRY FLOYD
January 1968

PART ONE

The People of Eastern Nigeria

1 Language and Ethnic Groups

INTRODUCTION

THE omnipresence of human beings is a fundamental fact of Eastern Nigeria's geography. Inhabiting the 29,400 square miles of land area are some 12,400,000 people[1]: men, women and especially children, of various languages, dialects and cultures. Under peacetime conditions, one is seldom out of sight of people and their habitations on journeys around the region; in few parts of Eastern Nigeria is it possible to feel isolated from a vibrant and emerging society. The reassuring sight of men and women about their business, walking or cycling along roads, forest paths and 'bush' trails in the countryside, or along streets in the villages and busy towns, is the reality of life in this significant section of Africa. Only over limited stretches, as in the remote interior of the Oban Hills or the swamps of the lower Niger Delta, are populations negligible or non-existent.[2] In these characteristics Eastern Nigeria is distinct from most of tropical Africa where sparser populations and extensive stretches of under-settled land are the rule rather than the exception.

The ubiquity of Eastern Nigerians and their general gregariousness have obviously had a profound effect upon natural landscapes. Vast stretches of the country are clothed in man-modified, secondary vegetation, interspersed with innumerable plots of growing crops or fallow land and criss-crossed by lines of communication: visible symbols of a vigorous social intercourse. Where populations are excessively (even dangerously) dense, as in parts of Owerri Province, the rural residences are wellnigh continuous. Thousands of villages, together with many small towns and

[1] According to the latest (and controversial) census of 1963, the total population of Nigeria in that year was 55,620,268. Eastern Nigeria with its 12,394,462 people thus contained almost a quarter (22%) of the population of the entire country, whereas the area of Eastern Nigeria constituted barely one-twelfth of the Republic's territory. Due to the Civil War, leading to the return home of over 2,000,000 Easterners, the population had swollen to approximately 14,500,000 by mid-1968, and famine was widespread.

[2] For the identification and location of geographical features and places named in the text, reference should be made to a good atlas.

growing urban areas across the region, provide homes for farmers, traders, businessmen, industrial workers and the new urban proletariat.

A further observable characteristic of the people of the Eastern Provinces, in addition to their profusion, is a remarkable range in their physiques; notably in stature, skin-colour and physiognomy or facial features. Quantitative measurements and analyses of these physical anthropological differences have not been made on any regional scale but a recent pioneer investigation into this important subject indicates that the diversities are pronounced.[1] While there are no differing racial groups among the indigenous population – they may all be classed broadly as Negro or Negroid – it is probable that various racial 'strains' contributed in former times to the physical make-up of the present inhabitants. Striking variations in pigmentation and height appear today not only between ethnic groupings or tribes but also within sub-tribes and clans. Thus while the majority of Ibo men tend to be short in stature, among Nsukka Ibos the incidence of tall and lean men is particularly noticeable. Skin-colour may range across Eastern Nigeria from black and very dark brown to light brown and even yellowish brown. Facial features may be fully Negroid, with flat noses and thick everted lips, or more Caucasoid in characteristics, with thinner noses and lips, and more angular faces. Tribal markings on the face and torso, once widely practised and a useful key for differentiating ethnic groups, are far less common among today's younger generation and may well be dying out.

When attire is considered, we have another colourful criterion with which to emphasise the impressive variety in the cultural make-up of Eastern Nigeria. It may not be feasible nowadays, except on traditional festive occasions, to differentiate tribal groups by their costume, since many people (particularly in the towns) have adopted Western-style dress for everyday wear: shirts and shorts, trousers or a smartly-tailored suit among the men; skirts, blouses and attractive frocks among the women. Alternatively, many Easterners wear Nigerian dress which is Hausa or Yoruba in origin, but which has become popular as a national costume. Nevertheless, the change from the traditional to Western modes of dress, which began with the Nigerians' earliest contacts with the white man, is by no means complete.

For back in Ibo villages, on farms, in smithies, on palm-trees and even in establishments where the Ibo is employed as a manual worker,

---

[1] P. A. Talbot and H. Mulhall, *The Physical Anthropology of Southern Nigeria. A Biometric Study in Statistical Method* (Cambridge, 1962).

one may still come across the work dress of the Ibo young man – a pant or a piece of cloth wrapped round the waist in which a matchet is worn. This scanty clothing becomes complete with an ivory armband and a necklace of black thread from which hangs a pendant of wild pig's teeth, which though a purely ornamental object, reflects lots of power accumulated in well-developed muscular chest and arms – power required for tilling the soil, for wrestling and for his other duties to his family.[1]

Among the women of the villages too, traditional clothes are still much in evidence: several yards of patterned print skilfully wrapped around the waist and reaching to the ankles, a blouse and headtie.[2] These costumes with their bright colours, like those of the men, are most becoming to dark faces in the strong sunlight. Young children in the rural areas sensibly run about with no clothes on (or, in the case of the girls, with a string of bright beads around the abdomen) until they are old enough for school uniforms.

HISTORICAL ORIGINS[3]

The generally accepted way of describing and differentiating the inhabitants of Eastern Nigeria is through the identification of linguistic groups or clusters of people sharing a common tongue or related dialects. These groups are commonly referred to as tribes, as in fact they are in the original sense of the term: social aggregations comprising a series of families, clans or generations and sharing similar characteristics and customs. In recent years the terms 'tribe' and 'tribalism' have come to be associated with emotional and political sentiments frequently of a derogatory nature, and disruptive of harmonious relationships between different peoples. For these reasons the word 'tribe' is today eschewed by many Nigerians.

The categorisation of peoples by language is a convenient academic device although, if it is followed too rigidly, it can obscure similarities in modes of living and geographically significant spatial variations in culture which cut across purely linguistic boundary lines. A number of distinct languages are spoken in Eastern Nigeria, such as Igbo, Ibibio, Efik, Ijaw, Ekuri-Yakurr and others; while most of these languages are more or less

[1] O. Nzekwu, 'Ibo People's Costumes', *Nigeria Magazine, 78* (Sept. 1963), p. 164.

[2] For an informative account of a traditional Eastern Nigerian cloth, Akwete (woven by women), see L. O. Ukeje, 'Weaving in Akwete', *Nigeria Magazine, 74* (Sept. 1962), pp. 32–41.

[3] A working paper by P. E. B. Inyang of the Department of Geography, University of Nigeria, has been used as a basis for the following sections.

distantly related to each other, linguists have shown that the differences among them are often profound rather than slight. Because of this diversity of tongues, intra-regional communication has at times presented difficulties; indeed the multiplicity of languages is still a divisive factor so far as regional coherence is concerned. To be sure, English is the recognised lingua franca but the bulk of the population still feel most at home in their own vernacular and it is doubtful if they will ever want to become entirely literate in a strange and alien language. Neither may we assume that the indigenous languages are dying out; on the contrary, the number of people who speak them is increasing with the rapid population growth. A bilingual pattern will probably remain for many years to come, with English as the language of officialdom and the business world, the local tongue as one of social intercourse and family life.

To enquire into the origins of the different language groups of Eastern Nigeria is to raise many more questions than can at present be satisfactorily answered. There is a great dearth of historical materials and reliable evidence, since local informants derive their knowledge largely from memory and unwritten traditions. These tend to be modified from one generation to the next under the influence of fresh experiences, leading to the creation of 'new' traditions.

There appears to be little oral tradition concerning the earliest population movements into the area which is now Eastern Nigeria. Discovery of Middle Stone Age artifacts (as recently as 1966 near Afikpo) indicate that people were dwelling in the region in prehistoric times, although it is impossible at the moment to date the beginnings of such Paleolithic settlements. The consensus of archaeologists and historians is that these peoples came from the north, at a period when the Saharan environment was more humid than it is today. East–west movements are believed to have come much later, with groups of shifting cultivators and hunters entering Eastern Nigeria from the Cameroon Highlands and Central Africa, and a counter-flow taking place from the west, perhaps around A.D. 700 to 1000. With the emergence of the Benin kingdom in the fifteenth century, immigration from west to east was intensified.

Influences from Northern Nigeria have also been felt in historic times. About A.D. 800, as a result of pressure from North Africa which the Arabs, flag-bearers of Islam, had invaded, desert tribes moved southwards to enter the territory of modern Nigeria, establishing such powerful states as the Empires of Kanem and Bornu. The political development in the north was not without repercussions in the forest zones of the south. In particular, invasions from the north into Nupe and Yoruba country led to population migrations into the forested areas east of the Niger River, prior to A.D. 1000.

THE IJAW

It is probably among the peoples of the Niger Delta (Yenagoa and Degema Provinces) that living descendants of the earliest inhabitants of the region are to be found. The language of the creeks and rivers areas is Ijaw or Ijo although there are many dialects from one section of the Delta to another, some so strong as to suggest that they may be the much-modified remnants of other language groups. The classic 'melting-pot' concept of linguistic intermixing has been proposed to explain the cultural situation in this swampy *milieu*, which has been a natural refuge over the centuries for tribes fleeing from domination by numerically stronger or better organised societies on the immediate mainland or in the interior.

Talbot has expressed the belief that Ijaw – classified as a Sudanic language – is the oldest of all Nigerian languages (of which there are more than 200) and is possibly one of the most ancient tongues in West Africa.[1] The very word 'Ijaw' is a corruption of 'Izon' (which the Ijaw call themselves) meaning 'truth'. It has been suggested that the Ijaw consider themselves the true or first inhabitants of Nigeria, as distinct from those who came later, who are not.

Among the Ijaw, traditions indicate that they are a very mixed people, with numerous clans and sectional differences. For example, the inhabitants of Brass claim to be descended from a group of Bini from Benin City who explored that part of the Delta around Brass River and settled on a small tract of land now known as Nembe. An alternative account of the origin of Nembe involves a royal army of the Oba of Benin, which elected to migrate to the Delta with its commander rather than face the vengeance of the king over the death of his eldest son in battle. There are other versions of the founding of Nembe-Brass and further Delta communities offered by other Ijaw-speaking peoples. A commonly-repeated theme is that of quarrels over the equable division of game killed in communal hunting expeditions, leading to wider dissension and eventual migration of the offended parties.

In the eastern sections of the Niger Delta, the people of Bonny and Okrika believe that their ancestors came from Iboland. West of Okrika and Port Harcourt is Abonnema, principal town of the Kalabari or New

[1] P. A. Talbot, *Peoples of Southern Nigeria* (London, 1926), vol. iv, pp. 72, 82.

Calabar people who claim to be an offshoot of the Efik of Old Calabar, whence they fled during a civil war. Today Bonny, Okrika and Kalabari tongues are classified as Ijaw dialects.

The Ijaw were the first Eastern Nigerians to have contact with Portuguese explorers in the fifteenth century and with subsequent European slave traders in the sixteenth and seventeenth centuries. When the slave trade was eventually replaced by legitimate trade in palm oil, they served as middlemen between the Whites and the people of the palm oil producing areas of the hinterland. Indeed the story of the palm oil trade is essentially the history of the growth of Ijaw ports such as Bonny, Okrika, Buguma, Abonnema, Degema, Brass, Nembe and Akassa, all within easy reach of the ocean-going vessels plying between Europe and the Guinea Coast. The historical and political roles of the Oil River States have been expertly evaluated by Dike and Jones.[1]

Today, the coastal Ijaw are essentially a breed of fishermen and traders by water, which is only to be expected from the nature of their homeland. In the central and northern portions of the Delta some farming is undertaken along the back slopes of river banks and elsewhere where the land is not swampy. The linear, compact, riverine villages of rectangular, stick-framed mud houses may contain many hundreds of people – communities such as Odi, Kaiama and Sabagreia in the northern Delta. Only occasionally in these villages are improved homes of concrete and pan roofs visible. The historic Ijaw towns support populations to be numbered in tens of thousands, e.g. Buguma, Abonnema, Nembe, Oloibiri and Amassoma. Nevertheless, for Eastern Nigeria as a whole, Ijaw-speaking peoples comprise little more than three per cent of the total population (Fig. 1.1).

Considerable numbers of Ijaw-speakers are to be found in mid-western Nigeria, west of the highly artificial boundary line in the Delta which delimits the Eastern and Mid-Western Provinces. A strong desire to establish a separate Ijaw-speaking Rivers State prevails in the area, based on historic sentiment and the fear of political and economic domination by numerically-superior groups in Eastern Nigeria.

THE IBIBIO-EFIK

Occupying the south-eastern section of the region, consisting essentially of the lower Cross River Basin and adjacent lowlands (Uyo and Southern

[1] K. O. Dike, *Trade and Politics in the Niger Delta 1832–1885* (Oxford, 1956); G. I. Jones, *The Trading States of the Oil Rivers. A Study of Political Development in Eastern Nigeria* (London, 1963).

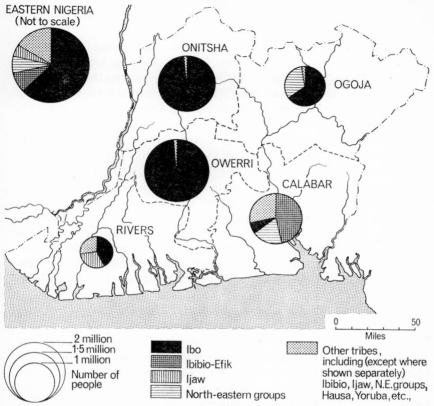

EASTERN NIGERIA
(Not to scale)

ONITSHA

OGOJA

OWERRI

CALABAR

RIVERS

0 ____ 50
Miles

2 million
1·5 million
1 million
Number of
people

Ibo
Ibibio-Efik
Ijaw
North-eastern groups

Other tribes,
including (except where
shown separately)
Ibibio, Ijaw, N.E.groups,
Hausa,Yoruba,etc.,

FIG. 1.1 *Percentage distribution of main tribal groups on a provincial basis,*
1953

Calabar Provinces), is the homeland of Ibibio- and Efik-speakers; next
to the Ibo this is the most numerous and influential group of Easterners,
although amounting to little more than nine per cent of the total population (Fig. 1.1). If the Annang are included with this group, however, the
proportion is raised to some fifteen per cent.

The origins of the Ibibio, like those of the Ijaw (and, for that matter,
the Ibo) are shrouded in mystery and obscured by conflicting traditions.
They are sometimes referred to as a semi-Bantu group, indicating their
linguistic affinity with the Bantu congeries of languages in Central, Eastern
and Southern Africa. It is most probable that they migrated to their
present area from the Cameroon Uplands to the east and, from the Cross

River estuary, moved upriver as well as westwards (inland, and along the coastal swamps) until checked by other tribes.

The Ibibio language includes many dialects, the principal ones being Ibibio proper (Uyo), Efik (Calabar), Annang, Enyong, Eket-Ibibio and Andoni or Ibeno. In addition to the main Ibibio-speakers there are other, smaller groups, each with differing versions of their historic relationship with the main members of the ethnic cluster. There are other peoples in this south-eastern section which are not Ibibio in affiliation, including the linguistically obscure Ogoni, Ejagham (Ekoi) and Ododop. Some of these groups represent relic minorities from former migrations through the area.

Efik is the literary language of the Ibibio cluster. The Efiks live mainly in and around Calabar and are thought to have travelled downsteam to their present locality. The main body of migrants settled at Creek Town on the Calabar River and from there Old Town and Duke Town were developed. The latter, being nearer the mouth of the river and a more convenient anchorage point for the early trading hulks, was destined to become an important commercial centre, the town of Calabar. Like the Ijaw, the coastal Ibibio profited greatly from slaving and, at a later date, the oil palm trade.

A curious and persistent belief among many Ibibio people is that they are related in some way to the Jews of Europe and Asia (this belief is also held by certain Ibos). Instances are cited of similarities between Ibibio and Hebrew in sentence structure, idioms, proverbs and word usage. Common social customs, such as certain forms of purification, have also been noted. Were small colonies of Jews deported from Portugal and established along this section of the Guinea coast in the sixteenth century, to be absorbed by the dominant Negro population in later years?

Today, this part of the region has the highest percentage of Christians in Eastern Nigeria and, relatively speaking, a commendable level of literacy (Table 1.1). An interesting feature of the distribution of religious affiliation is the high proportion of Christians in the coastal and riverain areas (Calabar, Yenagoa, Degema, Owerri Provinces), the result of over one hundred years of missionary effort, particularly by the Anglicans, Methodists, Presbyterians and Roman Catholics. Since Christian missions have also been carrying the burden of primary and secondary school education, patterns of literacy tend to follow those of the frequency of Christian churches.

The Ibibios and adjacent peoples are basically farmers and fishermen. Those living nearest to the sea, such as the Ogoni, Andoni, Effiat and Mbo depend very largely on fishing for their livelihood, their catch being smoked

TABLE 1.1

Literacy rates of population seven years of age and over,
Eastern Nigeria (1953)[1]

Province and Division	Standard of education and literacy (% of population seven years and over)	
	Std. II or higher	Others able to write in roman script
Calabar Province	12·7	8·1
Abak Division	9·0	6·8
Calabar Division	17·5	10·4
Eket Division	12·3	7·3
Enyong Division	12·1	5·8
Ikot Ekpene Division	10·3	7·6
Opobo Division	11·6	4·3
Uyo Division	16·6	12·9
Ogoja Province	3·0	2·1
Abakaliki Division	1·2	1·0
Afikpo Division	3·1	1·5
Ikom Division	10·7	5·5
Obubra Division	6·0	6·3
Ogoja Division	3·5	2·1
Onitsha Province	8·8	5·1
Awgu Division	5·2	3·3
Awka Division	10·0	3·8
Nsukka Division	3·4	4·6
Onitsha Division	13·8	6·5
Udi Division	9·9	5·5
Owerri Province	14·0	6·3
Aba Division	14·6	7·3
Bende Division	14·7	7·6
Okigwi Division	17·6	8·0
Orlu Division	9·8	3·4
Owerri Division	13·0	5·2
Rivers Province	12·9	5·2
Ahoada Division	11·3	5·5
Brass Division	11·8	6·0
Degema Division	15·2	6·2
Ogoni Division	5·3	3·3
Port Harcourt Division	37·6	5·2

and traded inland for farm products. The Oron, Okobo, Eket, and Eg-
wanga are seasonal fishermen, going to sea when work on the land is over

[1] Extracted from Department of Statistics, Nigeria, *Population Census of the Eastern
Region of Nigeria 1953* (Lagos, 1953), table 7, p. 40.

and returning to the farms in time for the next planting. The collection of palm fruits is a time-honoured occupation of the Ibibio; in former times most of the produce was collected from wild trees but sizeable acreages have now been planted with groves of improved trees.

The rural Ibibio live in fairly compact villages or nucleated settlements, divided into wards or sections; if of sufficient size, the wards may be physically separated and sometimes behave as distinct villages. Rectangular buildings are dominant, the few round houses being of recent introduction. The houses are fairly large and high towards southern Ibibioland but tend to get smaller as one moves northwards. They average two or three rooms, although occasionally a ten-roomed house occurs, usually arranged in the form of a compound, the buildings surrounding an inner courtyard. Traditionally, the homes are built around a bamboo frame which is mud-covered. The roof is made of palm stocks or bamboo, then covered with mats prepared from palm leaves. Increasingly visible are new 'builder's' houses in the rural areas, two-storied dwellings of plaster-covered cement blocks, attractively painted, and roofed with corrugated aluminium or asbestos-cement sheets. Ibibio towns such as Ikot Ekpene, Uyo, Abak, Oron and Calabar are expanding rapidly and modern Western-style architectural forms are popular for new business premises, private residences, schools and churches.

The Ibibio are noted for their artistic creativity and industry in the fashioning of such household items as cane chairs, raffia baskets and mats, and wood carvings. Bicyclists heavily laden with chairs and baskets are a common sight on the roads leading into towns such as Ikot Ekpene, as the products of a flourishing cottage industry move to market places and co-operative shops for disposal into a cash economy.

THE IBO

By far the largest language group in the Eastern Provinces of Nigeria consists of Igbo-speakers (Igbo is classified as one of the Kwa group of languages). The Ibo comprise something over sixty per cent of the total population and occupy a little over half the area of Eastern Nigeria (all of Owerri, Onitsha, Enugu and Umuahia Provinces, and the larger part of Abakaliki and Port Harcourt Provinces). Like the Ijaw, the Ibo are also to be found west of the boundary between the Eastern and Mid-Western Provinces (beyond the Niger River), where rural settlements of Igbo-speaking groups extend almost as far as Benin City.

It is a wellnigh impossible task to trace the history of the Ibo or the origin of their nomenclature. Both have been lost in the 'vicious circle' of traditions. Migration to their present home in Eastern Nigeria from a distant land must have taken place very many years ago. Since their domicile in the region, villages have begotten villages to the extent that the traditions of the offshoots have beclouded the original Ibo tradition of their beginnings.

Over the generations, nevertheless, one Igbo-speaking group, the Nri, has been held in great respect through Iboland. In the light of this it has been suggested that the Ibo perhaps originated from the Nri or that their original ancestors founded Nri several centuries ago. The town itself remains the centre of a cult connected with the installation of chiefs, purification and title-making. The priests and diviners of Nri are still widely travelled and sought after at important religious ceremonies and social functions in Ibo areas.

G. T. Basden, an Archdeacon of Onitsha, who spent thirty-five years working among the Ibo, wrote: 'All my attempts to trace the origin of the name "Ibo" have been unsuccessful. My most reliable informants have been able to offer no other alternative than that it is most probably an abbreviation of a longer name connected with an ancestor long since forgotten.'[1] Later Europeans, Forde and Jones, anthropologist and administrator respectively, declared: 'Before the advent of Europeans the Ibo had no common name and village groups were generally referred to by the name of a putative ancestral founder. The word "Ibo" has been used among the people themselves as a term of contempt by the Riverain Ibo (Oru) for their hinterland congeners. The term "Ibo" also applied at first to the Ibibio who were later distinguished as "Kwa Ibo", after the principal river in their country.'[2]

There should be no over-emphasis on the pejorative aspect of the term 'Ibo' however. The regional rivalries and local loyalties to be expected over so large a geographical area led to an insistence by specific Igbo-speaking groups on a closer identification than that of a mere common bond of language, i.e. a place-name prefix such as 'Onitsha Ibo' or 'Ndi Onitsha', to impress on foreigners their distinctiveness as a people within the broader linguistic group of Igbo.

Lack of widespread intercourse in earlier centuries, resulting largely from social, political and economic barriers, also a suspicion of unknown

---

[1] G. T. Basden, *Niger Ibos* (London, 1921), p. 11.
[2] C. Daryll Forde and G. I. Jones, *The Ibo and Ibibio Speaking Peoples of Southeastern Nigeria* (London, 1950), p. 5.

terrain beyond the immediate horizons, gave rise to many regional dialects. Only in more recent times, with the impact of the slave trade (involving widespread movements by Aro Ibos and their mercenaries), peaceful trading in agricultural products, the mobility of Nri high priests and Awka smiths, the establishment of a British Protectorate over Nigeria in 1900 and the subsequent educational work of European missionaries, has a more common vocabulary and phraseology tended to emerge.

Probably the seven most important dialects today are Owerri Igbo, Onitsha Igbo, Nsukka Igbo, Bende Igbo, Cross River Igbo, Aro Igbo and Western Igbo. Internal political boundaries for the Provinces and Divisions follow approximately the linguistic boundary lines since they were drawn up taking into account the geographical pattern of cultural groupings. The inclusion of Aro Igbo in the listing of important dialects is in recognition of the prominence of the Aro in former times, despite their numerical limitation. The Aro were an enterprising and ambitious people who responded to the European demand for slaves by throwing up a remarkable political organisation. With the assistance of professional fighters enlisted from neighbouring groups – the Abams, Eddas and 'Abikiris' – the Aro were able to dominate Iboland, establishing settlements in various parts of the region (some are still flourishing today), acting as trading agents with the coast, and issuing rough justice in the name of their much-feared oracle, the 'Long Juju'. Movement in Ibo country became much easier under Aro domination and legitimate trade was stimulated, although the price paid for this was the tragic one of slaving.

The characteristic attitudes of the contemporary Ibo have been summed up as those of:

> extreme aggressiveness and enterprising individualism found in conjunction with an equally striking degree of tolerance and gregariousness, which enables these people to co-operate and live peacefully together in over-enlarged village communities in the most over-populated area of Negro Africa, and in spite of the fact that most of the people of any consequence in their communities live in a state of adjourned and undetermined conflict with each other.[1]

The Ibo are, furthermore, a people with a great commercial bent. Their aggressive enthusiasm and success as traders have carried them to the four corners of Nigeria, and they have been likened to the Scots in their canny financial transactions and ability to turn the 'penny-penny' in their

---

[1] G. I. Jones, 'Dual Organization in Ibo Social Structure', *Africa* xix (1949), p. 150.

direction. The impact of their initiative and drive will be reflected in the geo-economic assessment of the region in Part III of this study.

The chief occupation of the Ibo continues to be farming, although a large number of men, forced out of the villages by land hunger, have sought alternative livelihoods in the towns of Eastern Nigeria and the nation at large, particularly the capital city of Lagos. One of the most recalcitrant problems being faced by Ibos is, in fact, that of severe overpopulation in certain parts of Iboland, particularly Orlu and Okigwi Divisions in northern Owerri Province. This critical issue provides a recurrent theme in the present geographical appraisal of Eastern Nigeria and is considered in some detail in Chapter 2. Apart from farming and trading, certain Igbo-speaking groups are noted for their handicrafts, e.g. the Awka and Nkwerre for iron tools and brass vessels (some intriguing bronzes have been unearthed at Awka), the Awka for wood-carving, the Ekulu and Abakaliki for earthenware.

The rural areas, particularly where densities of population are high, are replete with large villages of rectangular or square mud buildings thatched with palm leaves or, increasingly as one travels northwards, with grass. Improved living standards over the last decade, the result largely of remittances from 'sons abroad' in the towns, have resulted in the appearance of many improved residences of more complex architectural design and constructed of timber and cement blocks. The shiny reflection from new corrugated iron and aluminium roofing sheets is a common sight in most Ibo villages today. Intensive allotment-type farming in the inner, compound land around homes and along the roads of the villages is another distinguishing feature of the high-density Ibo areas. At the other end of the settlement spectrum, Ibo towns such as Enugu, Onitsha, Umuahia, Owerri, Aba and Port Harcourt are busy, expanding centres with a planned look about them and attractively laid out by Nigerian standards. Modern, colourfully decorated offices and residences are also appearing in greater numbers year by year and are a vital feature of the contemporary urban landscape.

NORTH-EASTERN LINGUISTIC GROUPS

The remaining tribes to be considered are those occupying territory in the north-eastern part of Eastern Nigeria, essentially within Ogoja Province, also northern and central Calabar Province.

The unravelling of linguistic associations in this area is a more formidable task than almost anywhere else in Eastern Nigeria. While the languages of the north-eastern tribes are referred to broadly as semi-Bantui there are strongly marked differences from group to group (and also within groups), to the point where two adjacent villages may speak dialects which are mutually unintelligible. Wave after wave of migrants must have entered this area over the centuries, probably from the east and north-east, with consequent areal confinement, fragmentation and assimilation of the earlier settlers by later tribes.

The principal ethnic divisions are the Ekuri-Yakurr, Nkembe (Mbembe), Boki, Yola, Ukelle, Obanliks, Ejagham (Ekoi) and the Tiv, who are found in much larger numbers in Benue Province across the border (or 'Munsh Wall') in Northern Nigeria. But there are many sub-groups in addition, making any accurate representation of linguistic groups on a spatial basis most difficult.

The Ekuri-Yakurr, with numerous minor tribes such as the Akunakuna and Bahumunu, live in the Middle Cross River basin, south of the river, in south-west Ogoja, south-east Abakaliki and northern Calabar Provinces. They are essentially farmers with areas of fertile soils to work and extensive forested areas to exploit under 'bush-fallow' techniques. They also fish and make canoes, while some are keen traders, carrying their surplus agricultural produce down the Cross River as far as Oron and Calabar.

Northern neighbours to the Ekuri are the Nkembe, whose home is also in the mid Cross River area, but largely to the north of the river, Their two most important towns are Obubra and Ofunatam at the con- fluence of an important tributary, the Awayon, with the main river. There are many sub-groups, such as the Oshopong, Adun, Okum, Iyalla, Ofom- bonga, Nta and Nselle. They also farm and fish, although it is remarkable to note that some of the riverain villages in this area lack even the most rudimentary knowledge of techniques for catching fish.

The Boki of Obudu Division, with sub-groups such as the Bete (Obudu) and Iyala, are capable farmers with extensive tracts of land at their dis- posal. One of the new farm settlements is located in this area and takes its name from the Boki tribe. They also hunt for game and undertake some metalwork and pottery. The dominant house-type in southern Bokiland is rectangular, with palm-mat roofs, but as one moves northwards an in- creasing number of square and round huts with grass roofs are noticeable.

The Bete occupy the piedmont areas of the Obudu Plateau. On the much-dissected plateau itself, some forty-five miles from Obudu town and at an elevation of over 5,000 feet, are four tribes: the Utanga, Bechebe, Balegeta and Bebi, whose tongues are entirely dissimilar to the Bete. While the first three of the plateau groups have languages which are mutually intelligible, none of them can understand the Bebi who share their mountainous homeland. The Obudu Plateau, like the Niger Delta, provides another excellent illustration of the natural protection offered by a difficult and inhospitable environment to weaker, less organised groups who might well lose their identity before the challenge of stronger tribes in a more favoured physical setting.

Within the limits of the Obudu Cattle Ranch, which covers some forty square miles of summit land on the plateau, the occasional remains of stone pit-forts raise intriguing questions concerning the former occupation of this natural fortress region. The stone foundations of former fortified villages, faintly reminiscent of Machu Picchu in the South American Andes, may also be identified on precipitous slopes of the bounding escarpment. Even today, in the vicinity of the Obudu Uplands and on the plateau itself, one observes house types distinctly different from those in the remainder of the country. In places, circular mud and wattle (or purely mud) huts are covered by very large, grass-thatched, conical roofs extending almost to the ground, so that one must stoop low to enter by a restricted door. Elsewhere, particularly among the Bechebe-Balegeta dwelling in the perpetually damp cloud forest *milieu* of the plateau, small square residences are entered by an equally confined entrance, the floor of the interior being several steps below the outside surface.

Many of the smaller language groups in the north-eastern sectors of Eastern Nigeria await more detailed study by social anthropologists, linguists and philologists. Needless to say, the impact of the West through missions and governmental authority during the Colonial period has been the least pronounced in this part of the region. This is borne out by the statistics on literacy (Table 1.1).

2 Distribution of Population

INTRODUCTION

MAN and land are the two key factors in the geographic equation. To most geographers, quite the most fundamental map to construct and to examine is that showing the distribution of population; such a map is a visual indicator of those very man/land spatial relationships that it is our special province to study.

The map of population distribution quickly arouses our academic inquisitiveness since, in revealing where and in what numbers man lives, it inevitably poses the question: why? It becomes then the task of the geographer to analyse and to try to explain how far the revealed distributional patterns may be due to the influence of physical, social, economic, historical or other agencies, working either singly or in combination. Factors which are usually brought to bear on the problem (and mapped on a comparable base-map to that used for population distribution) relate to landforms, climate, water supply, soil conditions, vegetational cover and diseases on the physical side; historical events, social organisation, economic conditions such as farming systems, trading and communications and political considerations on the human side.

Once the soundest explanation of the spread of population has been arrived at, there still remains the ultimate question: what of it? What is the *significance* of the existing areal distribution of its people to the life of a country, and particularly to its plans for economic and social advancement? How and in what ways (if at all) may the pattern of human settlement be modified in order to improve the prospect of a more rational utilisation of natural and human resources, and thus to enhance the viability of the state in a highly competitive world? Needless to say, these questions are by no means easy to answer, and it may be presumptuous of geographers to assume that they can provide the answers. At the same time, the application of geographical methods to resolving problems of this nature is an inescapable obligation if we are to justify the inclusion of our subject in the realms of higher education. We believe that the geographically-trained civil servant, administrator, planner and businessman may well have a decisive contribution to make to the future of emerging areas such as Eastern Nigeria.

The problems to be faced in producing accurate population maps in the developing countries are legion.[1] It is also apparent that maps purporting to show present distributions and densities are historical documents even before leaving the printing press. The number of people in a country and their precise location are never static; rather are they ever-changing and, with reference to the gross total, invariably increasing at a disturbing rate. An evaluation of the growth rate of the population of Eastern Nigeria is difficult since reliable statistics are lacking. Registration of births and deaths on a region-wide basis has never been carried out. Sample demographic censuses to ascertain vital statistics have been few. A surplus of women over men is known to exist, however, the female to male ratio for the adult population being around 115 women to 100 men for the region as a whole. Important variations in the sex ratio occur at provincial and divisional levels, reflecting the overall pattern of population distribution.

The surplus of females over males does not mean that there is a large number of unmarried women. Polygamy is still widespread, particularly in the rural areas, and this cultural tradition enables the society to absorb the surplus women into marital and family life; all women may therefore be regarded as potential mothers. Life expectancy is regrettably low. Rural health officers estimate the average life span of rural residents to be 40–45 years. Infant mortality is distressingly high; health authorities estimate the average infant mortality in rural areas to be 300–400 per 1000 live births. In the Delta area, it may be as high as 600 per 1000. The fertility rate of women is estimated at 6–7 births per woman, of whom 3–4 children may stay alive.

Some indication of the rate of change in population growth within Eastern Nigeria around 1960 is given in Fig 2.1. The map reflects the best efforts of a statistician, utilising admittedly meagre and unreliable figures, to estimate birth-rates on a divisional basis.[2] Even if one allows for a reasonable margin of error, the trend is clearly towards an explosive increase in population in the immediate future and, if the present growth

[1] For a broad review of these problems, see K. M. Barbour, *Population in Africa* (Ibadan, 1963), also R. Mansell Prothero, 'Population Maps and Mapping in Africa South of the Sahara', chap. 6 in K. M. Barbour and R. Mansell Prothero (eds.), *Essays on African Population* (London, 1961).

[2] O. K. Mitter, 'On Birth and Death Rates in Nigeria' (University of Nigeria, Nsukka, 1965) (MS.).

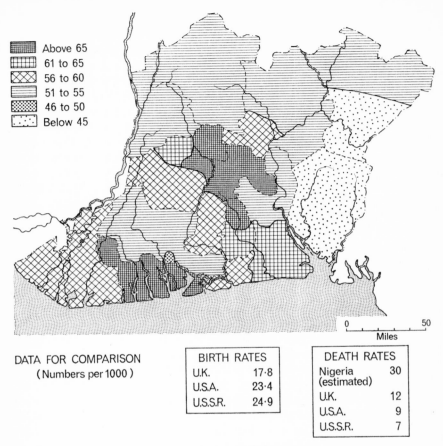

Above 65
61 to 65
56 to 60
51 to 55
46 to 50
Below 45

0   50
Miles

DATA FOR COMPARISON (Numbers per 1000)	BIRTH RATES		DEATH RATES	
	U.K.	17·8	Nigeria (estimated)	30
	U.S.A.	23·4	U.K.	12
	U.S.S.R.	24·9	U.S.A.	9
			U.S.S.R.	7

FIG. 2.1   *Estimated birth rates*, circa *1960* (*data from O. K. Mitter*)

rate continues, the prospect of a doubling of the number of Eastern
Nigerians within the next twenty to thirty years. The influx of refugees due
to the Civil War, and internal movements during the course of the conflict,
have inevitably modified the distributional patterns. Furthermore, it may
be many years before the intensified population pressures on the land are
alleviated by a new emigration of Eastern Nigerians to other parts of the
Federation of Nigeria. All the existing problems of the region deriving
from the size and inequable distribution of the present population will thus
be greatly exacerbated if adequate planning measures are not taken to
meet this potentially dangerous situation.

Census-taking in Eastern Nigeria, and indeed in Nigeria as a whole, has proved a formidable task on the several occasions that it has been attempted; the reasons for this are both too numerous and too involved to be recounted at this time. Suffice it to say that the survey of the characteristics of population in Eastern Nigeria which follows is inevitably circumscribed by the faulty and incomplete figures available for the study, despite the well-nigh heroic efforts of census teams in the field.

The 1953 census was the first attempt to carry out a house-by-house enumeration of the population throughout Nigeria. In addition to mere numbers, information was collected and tabulated on the distribution of the sexes in different age groups, principal occupations, main tribal groups, standards of education and literacy, religious affiliations, urban

TABLE 2.1

Inter-censal growth rate of population in Nigeria (1953–63)[1]

| Region | Total population | | Growth rate |
	1952–3	1963	%
Northern	17,007,377	29,758,875	75
Eastern	7,217,829	12,394,462	72
Western	4,595,801	10,265,846	123
Mid-Western	1,492,116	2,535,839	70
Lagos	271,800	665,246	—
TOTALS	30,584,923	55,620,268	85

populations, and data on the non-African population.[2] Since the detailed results of more recent censuses, in 1962 and 1963, have not been published, it has been necessary to make use of the 1953 figures for a number of the maps in this book.

[1] For a careful re-evaluation of the latest Nigerian Census figures, see C. Okonjo, 'A Preliminary Medium Estimate of the 1962 Mid-Year Population of Nigeria', in J. C. Caldwell and C. Okonjo (eds.), *The Population of Tropical Africa* (London, 1968), pp. 78–96. According to this study, the following figures represent the best estimate of the Nigeria Population in mid-1962:

Region	Total population
Northern	22,027,096
Eastern	12,332,046
Western	8,157,554
Mid-Western	2,365,091
Lagos	450,392
Total	45,332,179

[2] Nigeria, Department of Statistics, *Population Census of the Eastern Region of Nigeria 1953* (Lagos, 1953).

The latest Nigerian-wide census, in 1963, which was called for following the annulment of the 1962 figures, produced equally controversial figures for the nation as a whole, and was in part responsible for the rising tide of political agitation and unrest which culminated in the military *coup* in Nigeria early in 1966. Table 2.1 compares the 1953 and 1963 census figures for all Nigeria.

The best explanation for the phenomenal but highly improbable growth rates revealed in Table 2.1 is to assume an under-count at the time of the 1953 census, when Nigeria was still under British rule and the census was associated with taxation, land acquisition and military conscription; and an over-count at the time of the 1963 census, due to association of the count with voting registration and provision of amenities such as roads and piped water supplies to well-populated communities.

THE PATTERN OF POPULATION

The fundamental observation to be made is that Eastern Nigeria is one of the most densely settled areas in Sub-Saharan Africa. In some districts, in fact, it has rural densities rivalling those in the Nile Delta of North-east Africa which in turn are among the highest in the world for non-urban populations. According to the 1953 census the overall density for the region was 245 persons per square mile; the 1963 census results yield an arithmetic density of 420 persons per square mile. However these crude measurements disguise both the variable qualities of the land for supporting settlement and the very great contrasts in settlement densities which occur from place to place in Eastern Nigeria. Thus in 1953 Calabar Division in Calabar Province had an overall density of 50 persons per square mile while a section of Orlu Division in Owerri Province had a density in excess of 1,400 persons per square mile. In 1963, Calabar Division recorded 94 persons per square mile, while Orlu Division as a whole had an estimated density of 1,632 persons per square mile.

These density figures, it should be emphasised, represent a mainly rural population. The six large urban centres of Port Harcourt, Onitsha, Enugu, Aba, Buguma and Calabar have between them scarcely more than three-quarters of a million people, or 6·4 per cent of the total population. By adding the population of a further thirty-four towns listed in the 1963 census (although it is questionable whether they all fulfil a truly urban function), the number of town dwellers rises to a little over one and a half million, or 12·8 per cent of the total population (Table 2.2). Eastern Nigeria

TABLE 2.2

Urban centres of Eastern Nigeria, in rank order (1963)

	Town and Province	Population	Group
1	Port Harcourt (Port Harcourt)	179,563	Over 150,000
2	Onitsha (Onitsha)	163,032	,,
3	Enugu (Enugu)	138,457	100,000–150,000
4	Aba (Umuahia)	131,003	,,
5	Buguma (Degema)	100,628	,,
6	Calabar (Calabar)	76,410	75,000–100,000
7	Abonnema (Degema)	53,261	50,000–75,000
8	Awka (Onitsha)	48,725	25,000–50,000
9	Ugep (Ogoja)	44,945	,,
10	Ihiala (Onitsha)	40,198	,,
11	Ikot Ekpene (Annang)	38,107	,,
12	Afikpo (Abakaliki)	36,096	,,
13	Opobo (Uyo)	35,458	,,
14	Oron (Uyo)	34,163	,,
15	Abakaliki (Abakaliki)	31,177	,,
16	Bukana (Degema)	29,592	,,
17	Umuahia (Umuahia)	28,844	,,
18	Nsukka (Enugu)	26,206	,,
19	Owerri (Owerri)	26,017	,,
20	Nembe (Yenagoa)	25,032	,,
21	Okrika (Degema)	24,138	10,000–25,000
22	Omoku (Port Harcourt)	20,323	,,
23	Ozubulu (Onitsha)	20,000	,,
24	Oloibiri (Yenagoa)	19,314	,,
25	Awgu (Enugu)	18,504	,,
26	Amassoma (Yenagoa)	17,246	,,
27	Tombia (Degema)	16,462	,,
28	Oguta (Owerri)	15,310	,,
29	Uyo (Uyo)	14,470	,,
30	Ogoja (Ogoja)	13,694	,,
31	Ikom (Ogoja)	12,952	,,
32	Obudu (Ogoja)	12,633	,,
33	Brass (Yenagoa)	11,174	,,
34	Elele (Port Harcourt)	10,560	,,
35	Omoba (Umuahia)	9,914	Below 10,000
36	Aro (Umuahia)	8,800	,,
37	Bonny (Degema)	7,410	,,
38	Okigwi (Owerri)	7,266	,,
39	Orlu (Owerri)	7,234	,,
40	Ahoada (Port Harcourt)	6,329	,,

is therefore dominantly rural, with 87 per cent of her people living in hamlets and villages, or in isolated rural homes. By no means is this mass of the populace able to support itself by agriculture, as will be shown later. Many are engaged in petty trading and other secondary and tertiary occupations. It should also be observed that there is assuredly an inflation

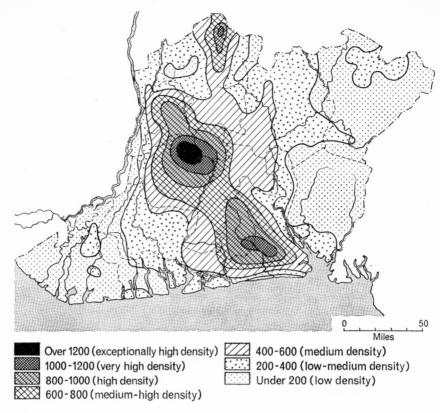

Over 1200 (exceptionally high density)   400-600 (medium density)
1000-1200 (very high density)   200-400 (low-medium density)
800-1000 (high density)   Under 200 (low density)
600-800 (medium-high density)

FIG. 2.2   *Density of population, 1963*
(*estimated number of persons per square mile*)

of the rural figures, and a deflation of the urban figures, arising from the 1963 census. Despite governmental urging that people should remain at their normal abode for counting, considerable numbers returned to their birthplaces from the towns (sometimes under the exhortation of their 'clan unions') in order to swell the local populace and so strengthen the case of their home village councils for additional amenities and welfare services.

Nevertheless the overwhelming character of Eastern Nigeria's population is that of rural occupancy and, considering the large numbers involved, represents a truly remarkable situation in view of widespread environmental handicaps: indifferent or poor soils, dissected terrain, seasonal inadequacies of water supply, difficulties of communication, and so forth, which are discussed more fully in Part II.

Examining more closely the distributional pattern of population, the most striking feature in Fig. 2.2 is the north-west to south-east axis of high densities running from the vicinity of Onitsha on the Niger to the Cross River estuary around Oron, a distance of some 120 miles and an average width of twenty-five to thirty miles. This belt of concentrated settlement has an average density of some 890 people per square mile, or over twice the regional average of 420 people per square mile. But this average again disguises significant variations along the axis, for at the northern end – the Ibo 'heartland' – are the densities of Orlu (1,632), Okigwi (1,267) and Awka (1,035) Divisions in Iboland (averaging 1,311 persons per square mile), and at the southern end, the densities of Abak (1,016), Uyo (1,171) and Eket (1,048) Divisions in Ibibioland (averaging 1,078 persons per square mile). Together the two cores of excessive density cover one-tenth (11 per cent) of the land area of Eastern Nigeria, and contain almost one-third (32 per cent) of the region's population.

The north-eastern and south-eastern core areas are separated by a wedge of comparatively lighter densities, those of Bende (516) and Aba (565) Divisions, representing the former 'no-man's-land' between Igbo- and Ibibio-speaking ethnic groups. The resultant pattern resembles dumb-bells or a weight-lifting device, with two heavy sections connected by a lighter piece.

On the flanks of the high density belt, population pressure on the land decreases fairly rapidly in most directions; the thinning-out process is exceptionally rapid northwards to the Anambra and westwards to the Niger flood plains from the northern core, also northwards up the Cross River and north-eastwards to the Oban Hills from the southern core. The steepness of the demographic gradient in these places raises challenging questions for the geographer.

Only to the far north of the region, in Nsukka Division (close to the border with the Northern Provinces of Nigeria), are rural densities to be found comparable to those along the main axis, and then only over a restricted area. The cluster of peoples around Enugu Ezike probably yields densities of over 1,000 people per square mile, despite lightly settled country both south-westwards and south-eastwards of the Nsukka Plateau.

Two zones of low population density occupy extensive areas to the east of the country: Calabar Province (the heavily forested Oban Hills) and

Ogoja Province (the upper Cross River Basin and Obudu Highlands); also to the south-west of the region, Yenagoa Province and the swamp-ridden Niger Delta.

In Ikom Division of Ogoja Province, the overall density is seventy-six persons per square mile and some stretches of country are virtually un-inhabited. Calabar Province shows an arithmetical density of ninety-four persons per square mile but most of the inhabitants are living in and around the town of Calabar in the extreme south, so that considerable tracts of territory are under-settled, forming the largest continuous stretch of sparsely populated land in the entire region. The Delta Province of Yenagoa has a density of some ninety-two persons to the square mile, but most of the Ijaw are dwellers in riverine villages and a sizeable proportion of each square mile is water or swamp rather than dry land, so that uninhabited if grossly inhospitable areas are again widespread.

The eastern and western low density areas together comprise almost one-third of the country (31 per cent) yet are inhabited by only one-tenth (10 per cent) of the region's inhabitants, a coincidental reversal of the man/land ratio of the high density zones.

Summarising the facts of rural population distribution in Eastern Nigeria, five density zones may be identified:

    (i) Areas of High to Very High Densities.

        These consist of (a) the axial belt running from south-eastern Onitsha Province through northern Owerri to Annang and southern Uyo Provinces, broken by a zone of medium density in Umuahia Province; and (b) northern Enugu Province (the northern part of Nsukka Division). The population exceeds 800 persons per square mile, and exceeds 1,200 persons per square mile in parts of Orlu and Okigwi Divisions of Owerri Province.

    (ii) Areas of Medium to High Densities.

        These flanking zones to the high density axis occur in southern Onitsha and southern Owerri Provinces, also in Opobo Division, south-west Uyo Province. Densities range from 600 to 800 people per square mile.

    (iii) Areas of Medium Density.

        These appear in a north-to-south line through the centre of the region, extending from Nsukka, Udi and Awgu Divisions in Enugu Province through south-western Abakaliki Province (Afikpo Division) to Bende and Aba Divisions of Umuhia Province, and Ogoni Division of Port Harcourt Province. Average densities are between 400 and 600 persons per square mile.

(iv) Areas of Low to Medium Densities.

These occur (a) on the eastern side of the medium-density zone, extending over most of Abakaliki Province and into Enyong Division of Uyo Province; also (b) as a southern, outer flank to the high axis belt, extending over most of Port Harcourt and Degema Provinces. Average densities range between 200 and 400 persons per square mile.

(v) Areas of Low Density.

These appear in the extreme eastern and south-western parts of Eastern Nigeria: Ogoja and Calabar Provinces, also Yenagoa Province. Densities are frequently below 100, but may extend to 200 persons per square mile.

FACTORS INFLUENCING THE DISTRIBUTION OF POPULATION

Consideration may now be given to those factors which have been responsible for shaping the pattern of population in Eastern Nigeria. For ease of presentation, these factors are grouped under the headings of historical, social, and environmental or physical. In reality, their influences have operated conjointly to produce the complex arrangement of human groups existing today in our study area.

HISTORICAL FACTORS

The present pattern of population is in large part a heritage of the events which have taken place in Eastern Nigeria in former times but particularly within the last 150 years.

In the first half of the nineteenth century the infamous slave trade was still flourishing, despite the rising tide of moral indignation in Western Europe and elsewhere at this long-established trafficking in human beings. Controlling the supply of slaves in this corner of Africa were the Aros, based at Arochuku in the lower Cross River Basin. Here was the site of the notorious shrine or 'Long Juju' which was the principal medium for recruiting slaves in Eastern Nigeria.[1] With the assistance of aggressive mercenaries such as the Abam and Edda, dwelling to the north of

[1] G. I. Jones, 'The Method of Obtaining Slaves in the Ibo Country and hence the Effect on the Population', *Journal of the Royal Anthropological Institute*, lxxix (1949). Also K. O. Dike, *Trade and Politics in the Niger Delta 1832–1885* (Oxford, 1956).

Arochuku, an effective network of settlements and lines of communication were established through which potential slaves could be recruited. Transported to Arochuku for the ostensible purpose of having their family problems and village disputes resolved by the 'Long Juju' oracle, defendants and often prosecutors alike were dispatched downriver to Calabar, one of the largest slave markets and ports in Nigeria, whence they were transhipped to the Americas. Alternatively they were marched overland, via Bende, a well-organised interior slave market in Aroland, thence to the port of Bonny on the Bight of Biafra coast, a strong rival to Calabar in the number of slaves exported yearly (some 20,000 slaves were shipped from Bonny in one year alone, 1790).

In this line of march lies one probable explanation for the persistence into the twentieth century of the distinct wedge of lighter population density between the two cores of the north-west to south-east high density axis. The Bende–Bonny slave-route reinforced in fact the prevailing pattern of settlement.

> The gap remained for long a frontier zone of outward expansion from the Ibo and Ibibio cultural centres, which may be identified with the very high density zones. As a nineteenth-century boundary zone between the Ibos of Aba and Bende Divisions, and the Annang groups of the Ibibio tribe, the gap remained an area of conflict particularly with the shortage of farmland about the turn of the century. Hence the relative low density of this 'shatter-zone'.[1]

The end result of the lengthy and extensive tapping of human resources from the middle and upper Cross River Basin was a massive depopulation of the inhabitants, from which these areas have yet to recover. Adding to the effects of this constant 'man-drain' were the ravages of disease and malnutrition which decimated surviving and scattered groups of the indigenous population still further.

With the final cessation of the slave trade and the pacification of Southern Nigeria by imperial authority at the start of the twentieth century, the pattern of population inherited from the pre-colonial period was to some extent stabilised. Large-scale regional movements of people under Aro surveillance ceased and across the countryside tribal groups sunk firmer roots in those areas they were occupying at the time of the British penetration. The 'Pax Britannica' brought added security and freedom from inter-tribal wars. The introduction of Western law and order meant as-

[1] R. K. Udo, 'Patterns of Population Distribution and Settlement in Eastern Nigeria', *The Nigerian Geographical Journal*, vi (Dec. 1963), p. 83. (Hereafter referred to as 'Population Distribution'.)

sured harvests from farmlands freed from the threat of raiders, hence improved nutrition, also the introduction of medical facilities (largely under missionary auspices) for combatting the fearful toll of endemic diseases. For these and other reasons a sharp increase in population growth occurred, the ramifications of which are so clearly evident today. A gradual but steady extension of settlements from established homelands, i.e. the creation of 'colonies' of new hamlets around parent villages has taken place in the intervening years, modifying but not radically changing the regional pattern of population derived from the nineteenth century.

One important exception to the 'freezing' of pre-existing patterns should be noted, resulting from the territorial acquisitions of European powers in this corner of Africa. As British and German administrators struggled to achieve the most favourable frontier for Nigeria and the Cameroons respectively, a displacement of population took place away from the zone of political and later military conflict. Deliberate destruction of villages is known to have taken place, undoubtedly with loss of life. This forced exodus may not have been on a large scale, for the area was sparsely peopled prior to the arrival of the white man, but it helps to explain why the border zone between Eastern Nigeria and the Cameroon Republic has remained virtually uninhabited down to the present time. Even British administration of the West Cameroon Mandated Territory in the interregnum following the First World War failed to encourage migration into this population vacuum.

SOCIAL AND CULTURAL FACTORS

The term 'culture' refers to those aspects of the life of a people which relate to their attitudes, objectives and technical abilities, i.e. their accustomed ways of living, their social organisation and their productive (or destructive) activities in seeking to nurture their society. Geographers are concerned with the spatial expression of these facets of culture within the homeland, visibly upon the landscape or as relevant pieces in the overall picture of the variable character of the region which geographers aspire to study.

For a fuller appreciation of the distribution of population in Eastern Nigeria, it is important to take into account those social forces which motivated the ancestral leaders of former communities, also the attitudes of the present generation of 'leaders of thought' whose decisions may induce changes in the present distributional pattern of the people.

For all inhabitants of Eastern Nigeria in former times, the cardinal considerations in the selection of areas for settlement were that they should contain defensible sites for villages and that they should offer land of sufficient size and quality to provide adequate sustenance for the group. The constant threat of attack from militant or revenge-seeking alien tribesmen, hungry neighbours whose crops had failed, or slave-raiding parties, was the inescapable reality of life up to the beginning of this century.

Hilltop defensive locations for villages and the protective environs of dissected plateau country or high plains are to be found in most areas of Eastern Nigeria where medium to high densities of population exist today. The prominent plateaux and flanking piedmont areas of the central and north-western parts of the region are among the most heavily settled sections; Onitsha and Owerri Provinces consist largely of undulating to rolling country with many ridges to offer defensible sites for communities. These watershed and interfluve locations were widely utilised in pre-colonial days and the resultant concentrations of population have persisted long after the need for the natural protection provided by landforms has disappeared.

Clear indication of the reliance placed on hilltop locations as defensive sites is provided by the many residual hills on the Nsukka Plateau, whose summits are crowned with the remains of former settlements and whose slopes are patterned with the 'fossil' remains of dry, stone wall terraces for the support of agriculture in the immediate vicinity of the settlements. This cultural overlay is thought to stem from the Igala occupation of the area in the eighteenth century. In many instances, these hilltop villages are still inhabited by the descendants of earlier occupants. Further south, at Maku in Awgu Division, there is an excellent illustration of hill settlements and terrace-type farming persisting to the present time.[1] The innate conservatism and attachment to the soil of rural peoples is a well-known social phenomenon which is amply illustrated in the Eastern Provinces of Nigeria.

By contrast, the open and flat countryside of the upper Cross River region (Abakaliki and western Ogoja Provinces) offered no natural protection for its inhabitants, and their subduing by the Aro-Abam fraternity was made the easier because of it. Another low density area, the Niger Delta, while substituting the protection of swamps for hills, has never drawn large numbers of people because of the acute lack of cultivable land. A further serious handicap to dwelling in the Delta was the wide-

[1] B. N. Floyd, 'Terrace Agriculture in Eastern Nigeria: the Case of Maku', *The Nigerian Geographical Journal*, vii (Dec. 1964), pp. 91–108.

spread danger of sleeping sickness, transmitted by tsetse flies which find their favoured habitat in the mangrove swamps and riverine areas.

As a precaution against depredations by Ibibio slave raiders and others travelling upriver from the coast, many tribes also shunned settlement along rivers and streams, despite their value as perennial sources of water. This avoidance of valleys reinforced the preference for interfluve locations for village sites.

> It is also possible that the tsetse had a hand in keeping people away from the valleys. The natives knew the symptoms of sleeping sickness though they never associated it with the fly. Nevertheless their instinct was sharp and once it was noticeable that the incidence of the disease was higher by the riverine groves, that was sufficient to initiate an antipathy to living near wooded valleys.[1]

The combination of efforts to avoid both predatory men and insects produced sparse populations along such water courses as the Niger (above the Delta), the Orashi, the Imo (running parallel to the Bende–Bonny slave route), the Kwa Ibo and the Cross River.

Tribes such as the Annang and Ibibio in the south-east, where hilly defensive sites were not available for settement, were obliged to utilise the high forest for protective purposes. Thus successive rings of closely spaced trees were left standing around communities, to act as natural walls of thick rainforest vegetation.

Turning to the requirements of farming land, the traditional system of agriculture was that of 'shifting cultivation', on which a voluminous literature has been produced. This simple yet effective method of producing foodstuffs by primitive cultivators meant that sizeable tracts of land were required by each community. Each year fresh clearings would be made in the high forest, and old plots allowed to revert to secondary forest cover, thus permitting regeneration of the soil under natural fallow conditions. At least seven to ten years of 'resting' were required before the land was fit for reclearing and replanting. In this way considerable stretches of country capable of producing the staple root crops were required for each settlement.

In its original form, shifting cultivation meant shifting habitation also, for when the land immediately adjacent to a village had been worked over, it was often simpler (if not expedient) to move the settlement to a new defensive site and to repeat the agricultural process in virgin forest. The consequence of this technique of farming on population spread was to

---

[1] Udo, 'Population Distribution', pp. 86–87.

promote rather low densities, perhaps in the order of twenty-five to thirty persons per square mile. As numbers grew, however (particularly in the Ibo and Ibibio core areas), and limits to tribal territory were imposed, so shifting cultivation gave way to rotational 'bush' fallowing techniques whereby the villages remained fixed in one location and a more stable and denser pattern of settlement ensued. Thus stability of settlement stands in direct relationship to the density of population.

In addition to the amounts of land required by different tribes and clans to maintain their rural economies, the quality of the land was equally important. Equipped only with cutlasses, short-handled hoes and other simple tilling tools, farmers naturally wished to avoid low-lying, ill-drained locations, with heavy clay soils and perhumid dense forests. The attraction of the uplands as protective sites for settlements was therefore under-scored by the relatively more attractive agricultural environment which they offered. The higher densities of population in the central and northern plateaux of Eastern Nigeria may therefore be understood also in terms of the traditional systems of rural land use of its peoples.

During the first half of the twentieth century, with increased pressure on the land, shifting cultivation gave way to 'bush' fallowing in most parts of the Eastern Provinces. Where found today in its original form, shifting cultivation exists only among small groups of north-eastern and eastern tribes and on the inner coastal plain in Ahoada Province, near the Niger Delta. Within the 'hearth' areas of the larger ethnic groups the intensifi-cation of land use has served to emphasise still further the disproportionate distribution of population over the region.

The spread of settlement within tribal territories was controlled by clan and family connections; groups of younger farmers moved away from their original homesteads to create the nucleii of new hamlets and villages, frequently named after the ancestral village or after the traditional heads of the family group (or their sons). The village of origin remained as the centre for social and religious gatherings and as a common market place. To maintain strong social connections between the parent villages and their 'colonies' required easy access by radiating footpaths and forest trails over the intervening countryside; the elevated watershed areas provided this facility, whereas low-lying, seasonally waterlogged land, swamps and rivers, impeded communications with other groups of the clan, as well as putting difficulties in the way of settlement and bush clearing.

Thus the social organisation of the past has persisted in the sedentary rural patterns of the present.

The paternal clan has evolved into the main village of a village group, the 'offspring' settlements have become 'quarters' or independent villages strongly bound to the parent village, in which economic and cultural institutions, market, church, school, maternity ward, oil mill etc., have been established.[1]

But beyond the horizons of the villages over the last half-century there have emerged new, non-rural communities engendering alien patterns of social organisation and requiring changes in the attitudes and objectives of those peoples who have become a part of the urban order. The growth of towns in Eastern Nigeria has been in direct response to the needs of a society emerging from an essentially internal subsistence economy to one of interdependence and external exchange with other countries and the world at large.

So far as the distributional pattern of population is concerned, the few prominent towns which have emerged to date are located in the medium and high density zones of the region, as one might expect, for their *raison d'être* is to meet the needs of the bulk of the population (the locational factors which helped to determine their precise positions are discussed elsewhere in this study). While attracting surplus population from the rural areas, they have done little to modify the basic pattern of population etched on the region from the nineteenth century.

Up to Nigeria's independence in 1960, the main activities of the towns were confined largely to administration and commerce. Government officials, the police, traders and teachers comprised the majority of the urban residents. Under the stimulus of development planning since independence, many new activities have been added to urban life, particularly at the principal 'growth points' of Port Harcourt, Enugu (the capital), Onitsha, Aba and Umuahia.

> While the administrative staff living in each town has been largely increased, commerce and trade have led to the establishing of larger shops or warehouses which employ clerks, salesmen and female workers, as well as office staff. Motorized transportation has led to the establishment of a transport business with repair shops, garages and filling stations; small crafts have developed in workshops; processing of agricultural products has been concentrated in larger plants in towns, especially oil and rice mills. The larger towns have attracted industries employing hundreds of workers in a single plant, the rising standard of living of a middle class in the towns has

---

[1] Government of Eastern Nigeria (Ministry of Works), *Master Plan for Urban and Rural Water Supply* (Enugu, 1962), pp. v–8. (Hereafter referred to as *Water Supply*.)

led to a demand for domestic help, and towns have thus grown far more than by the natural growth of population.[1]

A distinguishing feature of the urban inhabitants in Eastern Nigeria thus far has been the essentially temporary or transitory nature of their residence. Most town dwellers have been men (to a lesser extent women) whose chief social affiliations and loyalties have remained in their home villages. This special characteristic of the 'sons abroad' is a well-documented sociological phenomenon in other African countries also. Even if these migrants achieve prosperity in the towns they still consider themselves sons of their villages. They return home whenever possible, particularly on festive occasions, and contribute to community development projects sponsored by the village or clan unions. Invariably they build modern houses in their villages to testify to their social status, and to which they will eventually retire. These cement structures stand in marked contrast to the older and humbler mud homes of the permanent rural residents. It is debatable how long these ties to both town and country will survive, for there are already signs that the traditional obligations of rural society are becoming increasingly irksome to some town residents, particularly the new generation of young people who have been born, reared and educated in the urban areas. The urban populations may certainly be expected to continue their rapid growth, but the character of occupancy will probably change from one of 'sons abroad' to a more settled urban society.

Not all 'sons abroad' seek employment in the towns of the Eastern Provinces or elsewhere in Nigeria. Some are agricultural migrants, obtaining work on commercial plantations in the region or in Spanish-controlled Fernando Po. Some spread effect from the high density areas to low density areas in recent years is also indicated by the increasing number of migrant tenant farmers who may travel considerable distances from their homelands to rent land for cropping.[2] Sometimes, where negotiations can be successfully completed with the land-owning groups involved, whole communities may shift to new sites and assume a more permanent settlement. Parts of the lightly-settled Anambra Basin in the north west, also the northern Cross River area, are being filled in this manner. The Nike plains east of the Nsukka Plateau are illustrative of the process. Nike people have more land than they require as a result of their former activities as northern agents for the Aros. They raided far and wide with the intention of capturing slaves rather than acquiring land, but in the process

[1] Ibid., pp. v–4.
[2] R. K. Udo, 'The Migrant Tenant Farmer in Eastern Nigeria', *Africa*, vol. xxxiv, No. 4 (Oct. 1964), pp. 326–39.

they ended up with a sizeable extent of territory which became theirs by right of conquest. Today the Nike rent farms to land-hungry groups and have permitted the establishment of semi-permanent 'camps' of new settlers.

Nevertheless, these private contractual arrangements between different sections of the rural population have not had any marked impact upon the overall distributional pattern of population. The 'release-valve' effect has been scarcely discernible. Official Government encouragement for such transfers of surplus rural population on a greatly expanded scale might be of value, although this raises the thorny issues of the degree of coercion to be applied, and the complex social fabric of traditional tribal tenure which continues to act as a serious constraint against the more rational allocation of land resources in Eastern Nigeria.

PHYSICAL OR ENVIRONMENTAL FACTORS

Some indication of the role of the physical elements of Eastern Nigeria's geography in determining population distribution has appeared in the preceding sections. In particular, the significance of ridge-and-valley sections of the dissected plateaux and high plains as attractors of population, and of riverine areas and swamps as detractors of settlement, has been stressed.

Those aspects of the physical environment which are of special interest to geographers relate to geological formations (including mineral resources), water supplies, landforms, weather and climate, vegetational cover and soils. The systematic treatment of these topics occurs in Part II of this book, but it is appropriate to review certain features of the region's physical geography at this stage in so far as they aid our interpretation of the distribution of population map.

The larger part of Eastern Nigeria (eighty-five per cent) is underlain by sedimentary sandstones and shales of Cretaceous and Tertiary times. The eastern fifteen per cent consists of igneous and metamorphic rocks of Pre-Cambrian times. The landforms and soils of the region reflect this fundamental geological division. The ancient crystalline rocks of Ogoja and Calabar Provinces have produced rugged terrain and protective landscapes in the Obudu Highlands and Oban Hills: but steep slopes, heavy soils, excessive rain and dense forests, apart from historic factors, have inhibited settlement in these areas.

In parts of Enugu, Onitsha, Owerri and Umuahia Provinces, resistant Upper Coal Measure and False-Bedded Sandstone formations have provided the capping materials for the Nsukka–Udi–Okigwi plateau and escarpment; a further upland area westwards towards the Niger River consists of Nanka sandstones and related formations. Together, these elevated zones have offered abundant watershed and interfluve sites for Ibo communities. As clans increased in number, 'offshoot colonies', moving along lines of least resistance, found their easiest communication routes along the water divides, even along the crests of spurs in the case of the narrower interfluves. Areas of marked local relief, slopes deeply incised by valleys, ill-drained lowlands and rivers were the main obstacles to be avoided. Even at the present time, such adverse topographic features are utilised as boundaries between different clans and family groups.

Another physical factor which influenced the location and spread of population was the natural vegetation cover, in turn a reflection of drainage conditions, climate and soils. Temperatures and rainfall are uniformly high over all of Eastern Nigeria which has permitted the growth of evergreen tropical forests, although such natural forests tend to become higher and denser towards the southern reaches of the country, with shorter dry seasons, heavier rainfall and less well-drained soils. Clearing of the rainforest for purposes of shifting cultivation is not easy in the heavy clay soils of the eastern districts, also parts of the upper Cross River basin and the coastal lowlands. In areas of poorly-drained soil which becomes waterlogged in the rainy season, burning of trees and undergrowth is difficult; the clearing of sufficient land each year for farming would have been a formidable task.

On light, sandy and well-drained soils, on the other hand, clearing the original forest and later the secondary vegetation or 'bush' would have been a simpler process. Such soils were also easier to work with the implements available and were more suited to the staple root crops such as yams. To grow yams in low-lying heavy soils liable to waterlogging requires the laborious construction of large mounds to keep the tubers above the waterline, as any contemporary observer in Abakaliki Division will testify. It is understandable then that the free-draining sandy soils of the Plateau zones were favoured for agriculture, despite their limited inherent fertility, and provided a further inducement to settlement in these areas.

Apart from root crops, the oil palm thrives in sandy soils and palm products were to become a mainstay of the rural economy, particularly after the slave trade came to a welcome end. A further advantage of the sand lands lay in their suitability for road building. The intricate web of 'bush' tracks connecting villages, quarters and hamlets was readily convertible to motorable 'bush' roads with the advent of cars and lorries. No conversion was necessary at all for their use by the bicycle, now a ubiquitous and versatile carrier for a multitude of products. Free-draining dirt roads are passable even in the rainy season and require no costly foundations before tarring, unlike those traversing clayey soils. Along the watersheds, the need for bridges and culverts was kept to a minimum. Because of these physical advantages, and in response to the circulatory needs of the teeming population, the sandlands of Eastern Nigeria have acquired the densest network of roads in tropical Africa.

The only serious disadvantage to dwelling in the sandy uplands was, as it remains in many areas today, an acute shortage of surface water in the dry season. Owing to the porous nature of the soils there is little surface drainage and the water table is often too deep to be reached by shallow wells. Domestic needs are usually satisfied for the seven months of the wet season (April to October) but the water deficit over the remainder of the year has always presented a problem to the inhabitants and at times no doubt has limited the expansion of settlements. Catch pits and storage pots filled with evil-smelling contaminated water are still a common enough sight in Ibo villages on the sandlands. Fortunately, the hydrologic formations in the high density areas of Owerri and Uyo are capable of storing water and, through wells and boreholes, this has encouraged the growth and support of so large a population.

The historic concentration of communities on the sandlands is thus another key to comprehending the map of population distribution. One result of the lengthy occupancy of the sandy uplands has been a conspicuous change in the vegetational cover. Frequent burning and clearing of the forest by man eventually destroyed most of the large trees, removing the shade canopy from the lower plants and associated life forms in the ecosystem. The whole character of the vegetation changed in consequence. These changes have been most marked in the northern sections of Eastern Nigeria, where lower rainfall and longer dry seasons have reinforced man's efforts to check regeneration of the forest cover. The forests have thus been succeeded by tropical grasslands or man-induced 'derived savanna'. Only in protected locations, for instance around the villages and in sacred 'juju' groves, have relict patches of high forest been preserved. 'Bush' fallowing in the outer farmlands has now yielded to grass fallowing, and the seasonal task of clearing the land for planting is now much easier than in former times.

The price paid for this facility has been a high one. By disturbing the whole structure of the biotic environment, a marked deterioration in the capability of the sandy soils to produce domestic crops has occurred.

Never of high fertility, even under their original high forest cover, the now acid sands of the derived savannas and degraded forests to the south are severely impoverished and their continued utilisation under traditional farming methods can only bring about further lowering of yields leading to an eventual breakdown in productivity.

In the above analysis lies the answer to the extraordinary anomaly whereby the densest populations in Eastern Nigeria are supported at present on the poorest and least fertile soils in the region. This situation is entirely contrary to experience elsewhere in Africa and the world, where the heaviest concentrations of rural peoples occur on fertile alluvial or volcanic soils (e.g. the Nile Valley and Delta, the uplands of Rwanda and Burundi, the Ganges valley in India, the Java plains and the Hwang Ho lowlands in mainland China).

It is a situation which has most disturbing implications for the country; unless radical steps are taken to inject scientific farming techniques into the affected areas, and to offer alternative occupations to the great number of landless and jobless school-leavers, a mass exodus of peoples from the impoverished high-density belts will be unavoidable, changing drastically and irrevocably the distributional pattern of population of the last hundred years. This critical issue of the maldistribution of population in Eastern Nigeria will be a recurring topic in later sections of this study, and possible solutions to the dilemma will be suggested where possible.

3 Settlement Patterns

INTRODUCTION

WE should now take a closer look at the geographical patterns created by the association of residential buildings in Eastern Nigeria; first on a topographic scale when the arrangement of individual homes in hamlets, quarters and villages in the rural areas may be studied, then at the regional level when areas of different types of communities, dispersed or compact, continuous or isolated, etc., may be identified and mapped. It will also be necessary to examine briefly the settlement patterns of the major urban areas.

We have ascertained that some eighty-seven per cent of Eastern Nigerians continue to live in rural areas or, if employed in the towns, still have strong attachments to their home villages, own residences in them, and return for visits as often as possible.

The basic dwelling unit in the villages is the compound, comprising a closely-spaced group of small buildings constructed traditionally from sun-dried mud or mud blocks around a wooden frame, and roofed with thatching grass or palm fronds. These single-storied homes may be rectangular, square or circular in shape and contain one or at best only a few rooms. The compound may consist of several such houses, one for the head of the family, others for his wife or wives and their children, or for close relatives. Smaller structures are erected for storage purposes.

With rising expectations, higher prices for surplus foodstuffs, cash-cropping and the indispensible inflow of funds from the 'sons abroad', many compounds now contain larger, more modern, single-storied residences constructed of cement blocks, whitewashed or painted, and roofed with sheets of corrugated iron, aluminium, or asbestos-cement. More spacious, two-storied, 'builder's' homes are also increasingly in evidence: the country residences of successful migrants to the towns.

In addition to the dwelling quarters, the compound includes that part of the family's agricultural holdings which is under permanent cultivation, usually in the form of a vegetable garden in the shade of fruit trees.

This inner or compound farmland is in the hands of the women while the distant or outer farmland is traditionally the responsibility of the men. The entire compound of buildings, gardens and trees may range in size from under a quarter of an acre (especially in densely settled areas) to two acres or more where land has been more plentiful. A mud wall or wooden fence often surrounds the compound, an important defensive feature in earlier days, but, if still constructed today, largely to keep out stray animals such as goats and sheep. As a result of this spatial arrangement of man and land, discrete residences may be separated by distances ranging from twenty to two hundred yards; the contiguous nature of rural homes in the older villages of Britain and Western Europe, with a continuous row of cottages, is not found in Eastern Nigeria.

The hierarchy of rural settlements commences with the single isolated compound but, except in parts of the Upper Cross River Basin, this is not a common sight since the gregariousness of near-neighbours, as well as the security which stems from familiar faces at close quarters, are cherished by most Eastern Nigerians. In areas of nucleated settlement, the social outcasts, mentally-deranged, or former lepers are the most likely inhabitants of the remote compounds.

When a small number of compounds are grouped together, perhaps occupied by members of an extended family (kindred), the association is referred to as a hamlet or small village. Such hamlets may contain five to twenty compounds and thus shelter populations ranging from fifty to two hundred people. Hamlets are separated from other clusters of rural residences by their outer, rotational farmlands, oil palm 'bush', or stretches of secondary forest and savanna.

A group of hamlets occupied by members of a common clan forms the village proper, with populations varying from 1000 to 5000 inhabitants. In the sociological framework, 'the social structure of the village group is based on a theory of agnatic descent which regards each village group as a patri-clan, descendants of a common ancestor . . .'[1] The hamlets are located along paths or roads radiating from the common centre (the parent, ancestral community) where the churches, primary schools and market place are situated. The built-up sectors of a large village may cover three to five square miles, comprising the several clusters of hamlets and the core settlement. The perimeter of such a village is up to two miles from the centre although, with the irregular shape of most rural settlements, this distance is highly variable. The total area of compounds, outer farmlands and 'bush' may amount to fifteen square miles.

[1] G. I. Jones, 'Ibo Land Tenure', *Africa*, xix (1949), p. 311.

In favoured areas, Ibo villages have grown into super-settlement units or rural 'towns', with populations well in excess of five thousand, and total areas of twenty to twenty-five square miles. These huge concentrations comprise essentially agricultural peoples so that the title of 'town' is not strictly accurate, although in terms of sheer numbers it may be appropriate. The super-village is divided into several quarters, separated from each other by a belt of woodland, each quarter being made up of many hamlets and large compounds grouped along connecting paths. There are also many country homes of families residing semi-permanently in the urban areas. The main quarter or ancestral home is the only part of the settlement which may claim the functional attributes of a town, having urban activities in the form of stores, governmental agencies and other services.

In the absence of a detailed breakdown of population statistics for the 1963 census, the following rough estimate of the number of villages and their size in Eastern Nigeria has been made.

TABLE 3.1

Villages in Eastern Nigeria, 1963 (estimated)

		%
Total number of villages	6750	100
Small–medium-sized villages (less than 1000 inhabitants)	4000	59
Large villages (1000 to 5000 inhabitants)	2500	37
Super-villages (above 5000 inhabitants)	250	4

REGIONAL DISTRIBUTION OF SETTLEMENT TYPES

It is now possible to make some observations on the variable characteristics of rural settlements in Eastern Nigeria on a regional basis.

Graphic representation of the size and distribution of settlements according to the 1953 census is provided by Fig. 3.1.[1] The territorial extent of the hamlets, villages and super-villages reflects in no small measure the nature of the topography in the areas of settlement. Where level land prevails and no natural features impede the growth of the villages, a rural sprawl or overspill results, checked only by impingement upon the land of adjacent villages. Where, on the other hand, level land is at a premium

[1] The reader who is interested in rural settlement patterns at a topographic scale is advised to study selected 1/50,000 sheets of Eastern Nigeria, published by the Federal Surveys Department, Lagos.

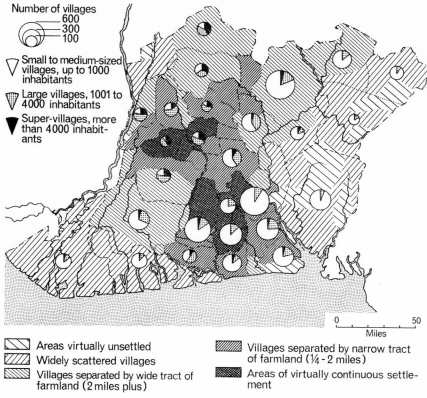

FIG. 3.1   *Size and distribution of settlements, 1953*

and the terrain is broken, villages along the hilltops and ridges tend to be more compact as the sites suitable for building are restricted. In low-lying swampy country, where dry land is scarce, villages are likewise strongly nucleated and limited in size.

Five zones of settlement types are identified, and these may profitably be compared with the five zones of density of population summarised on pages 42–3.

   (i) Areas of virtually continuous settlement.

  (ii) Areas with villages separated by a narrow extent of farmland (¼ mile to 2 miles).

 (iii) Areas with villages separated by a broad extent of farmland (over 2 miles).

 (iv) Areas of widely scattered villages.

  (v) Areas virtually unsettled.

(i) AREAS OF VIRTUALLY CONTINUOUS SETTLEMENT

In southern Onitsha Province, Orlu and Okigwi Divisions of Owerri Province, also Aba, Ikot Ekpene and Abak Divisions to the south-east, rural occupation of the land is virtually complete. In many sections, settlement is continuous with the dispersed compounds of one village adjoining those of the next. The establishment of so many homes has meant the virtual disappearance of outer farmland and the abandonment of 'bush' fallowing. The forest has been largely replaced by oil palms and fruit trees while food crops are produced in the compound plots. In the Igbo-speaking areas, the innumerable clusters of compounds may be identified with a more limited number of villages, whose populations are very large in consequence. In the Ibibio areas of heavy concentration, due to differing social and clan patterns, the same density of settlement is reflected in a large number of small hamlets and villages.

One noteworthy feature of residential patterns in the continuous-settlement zone results from the network of roads. The road construction programme has in fact had a vivid impact on settlement patterns in many parts of the country.[1] In order to share the benefits of easier contacts with more distant communities, particularly the towns, many people are today constructing houses by the roadside. New hamlets or quarters of older settlements sited away from the main roads are being developed, their modern linear spread in marked contradistinction to the traditional pattern of the historic nucleus. Awka, and Uzoakali in Bende Division provide good illustrations of this process. Along stretches of major roads in the densely settled areas, the succession of homes in ribbon-development fashion is most obvious. In places, the houses do not face the road directly, but have their frontage along a service road parallel to the main road, while the backs of the compounds confront the highway traveller. Elsewhere, due to a screen of 'bush' or planted trees, the parallel alignment of compounds is not detectable from the road. The setting-back and screening of houses from the road is presumably to offset the disturbing influence of lorries, cars and pedestrian passers-by.

(ii) AREAS WITH VILLAGES SEPARATED BY A NARROW EXTENT OF FARMLAND ($\frac{1}{4}$ MILE TO 2 MILES)

In areas adjacent to those with continuous settlement there is limited

---

[1] W. B. Morgan, 'Settlement Patterns of the Eastern Region of Nigeria', *The Nigerian Geographical Journal*, i (1957), p. 29.

outer cropland for farmers to utilise although the filling-in process is taking place rapidly.

In Iboland the pattern is again that of many dispersed but closely spaced hamlets and compounds, combining statistically to form large villages. The amount of open land remaining between settlements is very small. In Ibibioland, a larger number of small- to medium-sized villages prevails, with patches of farmland and 'bush' extending from one to two miles in most sections of Uyo, Eket and Opobo Divisions. Ribbon development of homes along motorable roads is another characteristic which these areas have in common with the continuously-settled zones.

(iii) AREAS WITH VILLAGES SEPARATED BY 2 MILES OR MORE OF FARMLAND

In Abakaliki and Western Ogoja Provinces, also northern Enugu, southern Owerri and Port Harcourt Provinces, the distribution of rural settlements is marked by clusters of farming peoples surrounded by sizeable areas of cropland, 'bush' and grass fallow, or secondary (in Port Harcourt Province, even original) high forest.

In the upper Cross River Basin, the recent resettlement of the area has led to an unusual number of isolated hamlets and even compounds, apparently unconnected with each other or with a parent community, set up without any clearly distinguishable pattern of clan affiliations. Some of these settlements represent the 'camps' of recent migrants to the area from the over-populated zones who have in fact no close social links with the land-owning communities. If all these clusters of population are counted as discrete settlements, then Abakaliki Division has the largest number of 'villages' of any Division in the Eastern Provinces. Only with the development of roads and motorable trails has a discernible concentration of settlements emerged, with larger villages and markets being established in ribbon-like fashion along lines of communication.

On the savanna-covered plateaux of the north-west, much greater concentrations of population are found, grouped together in large and super-villages, but still separated by broad stretches of open country. The name 'grassland towns' has been proposed for these large agglomerations of essentially rural residents.[1] This nomenclature is not entirely appropriate. On the one hand (as already observed) the majority of the inhabitants are – despite their numbers – engaged in farming activities, not in urban pur-

---

[1] W. B. Morgan, 'The "Grassland Towns" of the Eastern Region of Nigeria', *Transactions and Papers of the Institute of British Geographers*, xxiii (1957), pp. 213–24.

suits; on the other hand, the super-villages are not set directly in the open grassland but rather within fairly dense groves of man-modified forest or 'palm bush', islands of tree-shaded settlements set in a surrounding 'sea' of savanna.

In Udi and Nsukka Divisions of Enugu Province, the vast village has achieved its fullest expression (Fig. 3.1) and, unlike the teeming villages of Orlu and Okigwi Divisions to the south, may still extend areally and increase in population due to the relatively larger acreages of surrounding farm and fallow land. At the same time, increasingly urban characteristics may be expected to develop, with farming in the hands of those willing to adopt new and more intensive techniques of earning a living from the land, encouraging in turn the growth of secondary and tertiary service activities.

In southern Owerri and Ahoada Divisions, to the south-west, small- to medium-sized compact villages are spaced far enough apart to permit sufficient land to be left in fallow for their subsistence-farming populations. If a community begins to outgrow its territory, it is still possible in parts of Ahoada to move to a fresh site (in true shifting cultivation fashion), rather than to resort to more intensive farming techniques. On the other hand, these relatively lightly settled areas have attracted the attention of the Eastern Nigeria Development Corporation, and several plantation projects requiring large acreages of 'spare' land have been initiated in recent years, putting land out of reach of exploitation by would-be shifting cultivators in consequence.

Considering the excessive concentrations of people immediately to the north and east of this medium-low density zone, it is surprising that no marked immigration and establishment of settlements has taken place in Ahoada Division over the last few decades. The less well-drained, heavier soils, dense forests, and the conservative attitudes and suspicion of foreigners of the land-owning communities, have probably combined to thwart any large-scale influx of outsiders to the area.

(iv) AREAS OF WIDELY SCATTERED VILLAGES

The eastern stretches of the region contain small hamlets and villages, limited in number and far apart from each other. In the difficult topography of the Obudu Uplands and surrounding piedmont areas, settlements are compact and lightly populated. The upper Cross River Basin around Ikom provides more favourable terrain and soils for larger and somewhat more frequent communities. The Oban hills and most of the Niger Delta are also zones of small and widely-separated villages. The

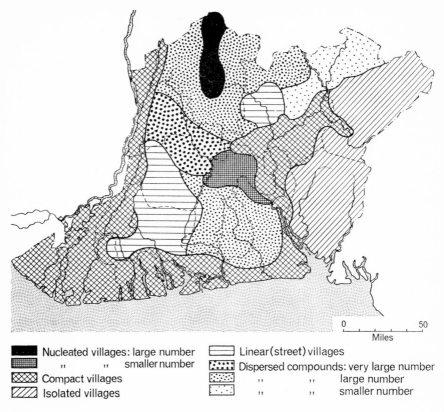

Nucleated villages: large number   Linear (street) villages
  ,,   ,,   smaller number   Dispersed compounds: very large number
Compact villages   ,,   ,,   large number
Isolated villages   ,,   ,,   smaller number

FIG. 3.2   *Structure of settlements (modified from R. Udo)*

yearly inundation due to the thirty-feet rise of the Niger means that secure settlement sites in the Delta are limited to a few areas of drier land along natural levees or on old river and beach terraces. Small linear hamlets and villages parallel to the distributaries or the coast are the rule, except in certain favoured locations both along the coast and inland where large villages and towns have developed, e.g. Brass, Buguma, Abonnema, Oloibiri and Amassoma.

(v) AREAS VIRTUALLY UNSETTLED

These areas contain a few very widely dispersed compounds and hamlets, and are to be found within and around the forest reserves of Calabar and

Ogoja Provinces, also in the Niger Delta. Their significance in the overall pattern of settlement types and distribution is minimal.

A further appreciation of the present-day structure of Eastern Nigerian villages is given in Fig. 3.2 which indicates those areas of the region where villages are either nucleated, compact, isolated or consisting of dispersed compounds. A careful analysis by Udo[1] confirms that the trend in recent years has been towards a disintegration of nucleated settlements and an increase in the number of dispersed if closely-spaced compounds, particularly in the high density areas where farmland is at a premium. The almost continuous compounds of southern Onitsha and Ikot Ekpene thus represent the final stage in the break-up of former nucleated settlements from the pre-colonial era.

URBAN HOUSING AND SETTLEMENT PATTERNS

The principal towns of Eastern Nigeria are modern creations, the product of the colonial period and subsequent politico-economic developments. Of the larger urban areas, only Calabar and Onitsha have developed from historical nuclei which are still evident in the twisting narrow streets and old-fashioned business premises and residences in the 'down-town' districts. Elsewhere, as in the case of Port Harcourt, Enugu and Aba, also such smaller towns as Abakaliki, Umuahia and Owerri, Western street patterns are dominant.

Where the terrain permits, a rectangular 'gridiron' plan has been imposed with straight roads crossing at rightangles. At other sites, straight or curving roads radiate from central places. The newer sections of indigenous towns such as Onitsha, Calabar, Awka, Ugep, Ihiala, Ikot Ekpene and Afikpo have also incorporated Western town planning. In the case of Ugep (in southern Abakaliki Province), the initiative of a native son has led to the adoption of a novel layout in the form of concentric ring-roads crossed by straightish streets radiating from several cores.

The urban, non-agricultural characteristics of these centres are indicated by their buildings: a central business district of shops, stores, petrol stations, administrative offices and service centres such as post offices and banks; a lorry park and large market with permanent, improved stalls; schools and churches; low-density, upper-income residential areas (often the old Government reservations for European officials) and the medium-

[1] R. K. Udo, 'Disintegration of Nucleated Settlement in Eastern Nigeria', *Geographical Review*, vol. lv, No. 1 (Jan. 1965), pp. 53–67.

high density, lower-income residential districts for the urban labour force of office and factory workers. Industrial layouts, where they exist, are usually on the outskirts of the towns (a functional treatment of towns is given in Part III, Chapter 16).

Business buildings range from old two- or three-storied edifices, with residential quarters above the shops for managerial staff, to large, modern, air-conditioned blocks reflecting Italian and Israeli rather than British colonial architecture. The same is true for the upper-income residential districts, whose houses of different styles and eras are surrounded by spacious gardens. In the lower-income districts, by contrast, buildings are contiguous and comprise small, single-storied dwelling places frequently filled beyond capacity with the official occupier and his family, also job-seeking relatives and friends visiting from the rural areas. The congested townships and crowded villages are twin manifestations of the over-population of parts of Eastern Nigeria.

1   *Derived savanna or woodland-savanna mosaic, Nsukka Plateau, 1966*

4 *Catastrophic soil erosion near villages of Agula, Nanka and Oko in Awka Division, 1966*

Above left: *Oil palm 'bush', with clearings for agriculture, Nsukka Plateau, 1965*
Below left: *Mangrove swamp forest, Cross River, near Calabar, 1964*

5　*Traditional Ibo residence set in oil palm 'bush', Nrobo village, Nsukka Division*

6　*Nucleated settlement of traditional Ekuri-Yakurr houses, Oderiga village, Obubra Division, 1966*

7 *A poignant scene: the queue for water from a standpipe near Nsukka, 1966*

8 *Ibibio fisherman's dwelling, Cross River, near Oron, 1964*

9   *An Ibo farmer from Nrobo village, Nsukka Division*

0 *An Ibo woman engaged in pot-making. The pot is constructed without the aid of a potter's wheel, 1965*

11 ‘*Egwu-Aja*’ *dance group of Ibo girls from a Training College entered in Annual Festival of Arts Competition held in Enugu*

12 *Ceremonial dance group of Ibo men from Onitsha, 1964*

13   *The new generation of young Nigerians. Graduating students of the University of Nigeria, Nsukka, at Commencement ceremonies, 1965.*

14   *Panoramic view of the University of Nigeria, Nsukka, 1965*

15 *The slash-and-burn methods of shifting cultivation under traditional agriculture.*
A plot of cocoyams in the Obudu Uplands, 1965

16   *Very large yam mounds in the hydromorphic soils of Abakaliki Province, 1963*

17   *A Compound plot of yams; the mounds with a top-dressing of wood ash. Nsukku Division, 1966*

19   *A Compound plot of cocoyams, Nrobo village, Nsukka Division, 1965*

20   *The fruits of the harvest. A well-stocked yam barn at Igbariam Farm Settlement*

18   Left: *A Compound plot of mature yams, firmly staked on poles, Nrobo village, Nsukka Division, 1965*

21 *Settlers preparing a citrus nursery and budwood garden, Igbariam Farm Settlement*

22 *Workers removing oil palm from a nursery for transplanting in main plantation blocks. Igbariam Farm Settlement.*

23 *Improved NIFOR variety of dwarf oil palm planted out on community plantations, ENDC estates and farm settlements*

24 *United States Agency for International Development machinery at work on Uzo-Uwani Farm Settlement*

25   *Settlers in the transit camp dining-hall, Boki Farm Settlement*

26   *Model village, comprising compact neighbourhood dwelling-units on 0·5-acre plots, Igbariam Farm Settlement*

27 *Preparing the land for rubber trees, uncleared high forest in the background, Dunlop Rubber Estate, Calabar Province*

28 *Young Gudali (Zebu) steer at ENDC Obudu Ranch*

29   *Hand-felling of tropical timber in Oban Forest. Observe the cutting-platform to raise axemen above the wing-like buttresses of the tree*

30   *Laborious, traditional pit-sawing method of cutting timber into planks,
near Obubra, 1966*

33   *Rotary cutter, Cross River Mill, near Obubra (Brandler & Rylke Ltd.), 1966*

31   Above left: *Mechanical transportation of felled trees to the landing-ground, to be moved to the sawmill by heavy duty trucks, 1966*

32   Below left: *Cross River Mill, near Obubra (Brandler & Rylke Ltd.), 1966*

34 *Exploratory drilling-rig in mangrove forest of the Niger Delta, 1964*

35 *Overcoming the problem of communication in the Niger Delta region: a helicopter service, 1964*

36   *Bonny terminal on the Bonny River, whence ocean tankers remove crude oil for shipment to refineries in western Europe, 1965*

37   *Aerial view of Alesa-Eleme refinery, shortly before going into production in November, 1965*

40   *Nigerian Cement Company* (*Nigercem*) *plant at Nkalagu, Abakaliki Province*

38   Above left: *Trestle bridge* (*over Okrika River*) *carrying the products pipeline from Alesa-Eleme refinery to the marine loading jetty on the Bonny River*

39   Below left: *Conveyor belt for laminations of asbestos-cement at Turner Asbestos Cement* (*Nigeria*) *Ltd., Emene*

41 *Workers at the Nigerian Glass Company Ltd., Port Harcourt. Grading finished products (bottles) for packaging*

42 *Reinforcing rods – products of the Nigersteel rolling-mill at Emene – being stacked for transshipment to dealers*

43 *Panoramic view of Enugu, capital of Eastern Nigeria, from Milliken Hill (to the west of the town)*

44 *New and old buildings intermixed along Ogui Road, Uwani Layout, Enugu, 1964*

45   *The Eastern Nigeria House of Assembly, Enugu, 1966*

46   *The ultra-modern Hotel Presidential in Independence Layout, Enugu, 1966*

47 *Canoes for river transport, Cross River, near Oderiga village, Obubra Division, 1966*

48 *The waterfront at Itu on Enyong Creek, Uyo Province, 1964*

49   *The harbour facilities of Port Harcourt, forty-one miles inland up the Bonny River, 1964*

50 *Road transport, revealing the popularity of bicycles, University of Nigeria, Nsukka, 1964*

51 *Lorries or 'Mammy Waggons' in the Aba Motor Park*

52   *The trunk 'A' road (A 11) linking Onitsha and Enugu*

53   *Completion of new bridge on the Ediba–Obubra road, Obubra Division, 1966*

PART TWO

The Physical Setting of Eastern Nigeria

Further Considerations of the Natural Environment

4 General Aspects and Geology

INTRODUCTION

THE rhomboidal territory of Eastern Nigeria covers some 29,484 square miles, or 8.3 per cent of the entire Republic of Nigeria (356,669 square miles), yet it supports 22 per cent or almost a quarter of the total population of the country. With a marked diversity of physical features, natural resources (both organic and inorganic), cultural groupings and settlement patterns, Eastern Nigeria in a sense epitomises the entire nation although it possesses certain geographical characteristics which are unique.

The land-mass of Eastern Nigeria extends from 4° 15′ N. to 7° 05′ N., and from 5° 32′ E. to 9° 16′ E.; it thus occupies very much of a tropical, indeed a humid tropical, location. For three of its four sides, the physical boundaries of the region are reasonably well defined. The 250-mile shoreline along the Bight of Biafra (Gulf of Guinea), while it comprises a confused complex of sandy, barrier beaches, lagoons, creeks and swamps, nevertheless forms a clear-cut southern boundary to the Eastern Provinces. On the west, the great Niger River and its distributaries, culminating in the vast and intricate drainage network of the Niger Delta, serve to delimit the boundary between Eastern Nigeria and the Mid-West State. In actual fact, the main channel of the Niger is utilised for only the first 60 miles of the 230-mile western boundary of the region. Some thirty miles south of Onitsha, the line leaves the river and, for a further thirty miles, lies on the average five miles to the east. The border then follows a highly irregular course, first to the Nun, then to the Forcados, before executing a number of right-angled bends and other abrupt changes of direction to reach the coast at the Pennington River. For most of its length, but particularly in the southern sector, the boundary has never been effectively demarcated; arising out of a series of agreements and concessions in former times, which tried to take into account local differences of physical and human geography (yet Ijaw and Ibo peoples are to be found widely on either side of the line), the European-imposed partition of the Niger Delta has become increasingly anachronistic.[1]

[1] For an authoritative study of the Niger River boundary, see J. R. V. Prescott, 'Nigeria's Regional Boundary Problems', *Geographical Review*, xlix (1959), pp. 485–505.

With the creation of the Mid-West Region in 1963 and frequent new discoveries of oil in areas traversed by the existing ill-defined boundary, the importance of a clearer demarcation is obvious.

To the east, the frontier between Eastern Nigeria and the Cameroon Republic follows the course of the Akpa Yafe river northwards from its confluence with the broad embayment of the Cross River estuary. Thence it traverses the sparsely-settled Oban hills, a low westward extension of an arm of the Cameroon Uplands. Descending from the plateau, the boundary runs with the upper Cross River for approximately ten miles before proceeding north-eastwards to the pronounced massif of the Obudu Plateau.

It is along the boundary between Eastern and Northern Nigeria that political demarcation has been least satisfactory in terms of coincidence with either topographic, hydrographic or ethnic divisions. The existing boundary was arrived at through a number of adjustments and concessions in former times in an effort to resolve persistent land disputes and conflicting political aspirations. Even today, sizeable Tiv communities may be found south of the line within Eastern Nigeria, south of the so-called 'Munshi Wall'. The boundary pursues an erratic course eastwards from the Niger, with northward, southward, even westward trends in places, before attaining the northern flanks of the Obudu Uplands. The rationalisation and clearer demarcation of the boundary between the Eastern and Northern Provinces of Nigeria is a task which awaits attention in the near future.

GEOLOGY

An understanding of Eastern Nigeria's physical features, in terms of landforms, relief, drainage patterns and soils, requires an introduction to the region's geology. Similarly, a fuller appreciation of human settlement patterns and economic activities (especially agriculture, mining and quarrying) is achieved through a knowledge of the rocks of Eastern Nigeria.[1]

Fig. 4.1 illustrates the geology in outcrop over the Eastern Provinces. A basement of ancient crystalline rocks – the Basement Complex – consisting of metamorphic rocks with intrusive granitic bodies, is overlain by a series of much younger Upper Cretaceous and Tertiary sedimentary strata, both marine and continental in origin. The sedimentary rocks were laid down 'either in a great embayment of the sea which reached north and east along the approximate lines of the present Niger and Benue Val-

[1] The author is indebted to R. A. Monkhouse (formerly of the Geology Department, University of Nigeria) for his assistance with this section of the book.

leys, or in the lacustrine or terrestial environments left by the retreat of the sea southwards into the present Bight of Benin'.[1] The sedimentary materials are largely derived from sub-aerial denudation and erosion of the rocks of the Basement Complex, with possibly smaller contributions from the sedimentary areas bounding the banks of the Niger beyond the territory of present-day Nigeria.

Igneous activity from the beginning of the Tertiary period onwards produced intrusive dykes and sills of doleritic and basaltic rocks and also gave rise to volcanoes. There is evidence of the latter in the eroded volcanic plugs around Abakaliki, together with beds of tuff, while more recent occurrences may be seen in the volcanic mountain ranges of the Cameroon massif.

Anticlinal warping has occurred at various times along a north-east to south-west axis which runs across the centre of Eastern Nigeria, from Abakaliki to Okigwi. This line is known as the Abakaliki Uplift; the crest of the anticline has been removed by secular erosion. Flanking the upfold to the south-east is a downfold, referred to as the Afikpo Syncline, running from Afikpo to Bansara. To the north-west of the region, there is a further flanking synclinal axis trending north-north-east to south-south west, known as the Anambra Shelf. Late downwarping of the Anambra Shelf is probably responsible for the large subsidence of the Niger Delta area during the Quaternary period.

The following geological formations and groups of rocks are recognised in Eastern Nigeria:

TABLE 4.1

Geological formations and groups of rocks in Eastern Nigeria

Formation	Group	Age
(a) Niger and Cross River Alluvium	i Delta and Alluvium	Quaternary
(b) Delta Formation		
(c) Coastal Plains Sands	ii Coastal Plains Sands	
(d) Lignite Formation	iii Bende-Ameki	Tertiary
(e) Bende-Ameki Formation		
(f) Imo Clay-Shales Formation	iv Imo Clay-Shales	
(g) Upper Coal Measures	v Plateau and Escarpment	
(h) False-Bedded Sandstones		
(i) Lower Coal Measures		
(j) Nkporo Shales		Cretaceous
(k) Awgu-Ndeaboh Shales	vi Cross River Plain Group	
(l) Eze-Aku Shales		
(m) Asu River Group		
(n) Basement Complex	vii Basement	Largely Pre-Cambrian.

[1] R. A. Monkhouse, *The Geology of Eastern Nigeria* (Nsukka, 1966), p. 1. (Hereafter referred to as *Eastern Nigeria*.)

A more detailed succession of rock formations with brief lithological and topographical summaries is given in the Appendix.

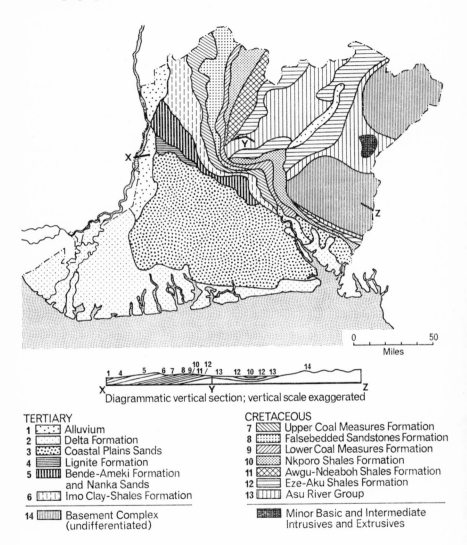

Diagrammatic vertical section; vertical scale exaggerated

TERTIARY
1 Alluvium
2 Delta Formation
3 Coastal Plains Sands
4 Lignite Formation
5 Bende-Ameki Formation
   and Nanka Sands
6 Imo Clay-Shales Formation

CRETACEOUS
7 Upper Coal Measures Formation
8 Falsebedded Sandstones Formation
9 Lower Coal Measures Formation
10 Nkporo Shales Formation
11 Awgu-Ndeaboh Shales Formation
12 Eze-Aku Shales Formation
13 Asu River Group

14 Basement Complex
   (undifferentiated)

Minor Basic and Intermediate
Intrusives and Extrusives

FIG. 4.1   *Simplified geology (after R. Monkhouse)*

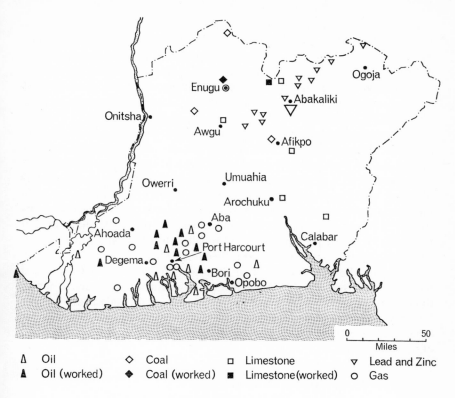

Oil △   Coal ◇   Limestone ▫   Lead and Zinc ▽
Oil (worked) ▲   Coal (worked) ◆   Limestone (worked) ■   Gas ○

FIG. 4.2   *Mineral deposits and power resources*

(i) DELTA AND ALLUVIUM GROUP

(a) Niger and Cross River Alluvium
(b) Delta Formation

The classical-shaped Niger Delta extends southwards from an apex near Aboh, some fifty miles south-south-west of Onitsha, and lies within both Eastern and Mid-Western Nigeria. Its easternmost limit reaches the coast near Opobo (Fig. 4.1) where it is probably contiguous with the western extremity of the far less impressive Cross River Delta. Together, approximately 5,000 square miles (18 per cent of the region) is floored by Quaternary deltaic deposits of the two principal rivers of Eastern Nigeria.

The Niger Delta (which is structurally very similar to that of the Mississippi) is built up of unconsolidated beds of predominantly coarse sands and gravels; there are also dispersed layers of silty clay and peaty matter, the decomposed remains of vegetation. The sediments near to the surface dip seawards at a low angle, being the topset beds in the present phase of development of the delta. At great depths, petroleum-bearing formations were located by the Shell-BP Petroleum Development Company in 1956 and are now being exploited by a number of other oil companies besides Shell-BP (Part III, Chapter 15).

The accumulation of materials over the last 75,000 years has been considerable. Deep boreholes sunk in the course of oil exploration have penetrated to depths of 10,000 feet without reaching pre-Recent rocks in the central part of the delta. Sedimentation continues at an exceedingly high rate at the present day; estimates at Onitsha of the amount of sediments transported annually by the Niger (average of yearly measurements from 1925 to 1957) indicate the movement of 660×10^6 cubic feet of material, of which 85·5 per cent is so-called 'wash load', 9 per cent is 'suspension load', and 5·5 per cent is 'bed load'.[1]

North of Aboh on the Niger, a zone of alluvial deposits (plains and terraces) often ten miles or more in extent, and several hundreds of feet thick, flanks the river, obscuring the canyon-like, steep-sided valley of the Niger which existed before the post-glacial rise in sea level. Similar although less extensive alluvial deposits are found around the flood-plain of the Cross River.

(ii) COASTAL PLAINS SANDS GROUP

(*c*) *Coastal Plains Sands.* The sedimentary rocks of the Coastal Plains outcrop over a very wide area, almost 25 per cent of the region, and are thus the most extensive surface formations in Eastern Nigeria. They constitute a broad plain sloping gradually towards the sea but northwards they rise to form low hills with elevations of 600 feet or more above sea level.

Coastal Sands deposits provide the most important aquifer or water-bearing body in the Eastern Provinces, a factor which accounts in part for the very high densities of rural population in the northern reaches of the Coastal Plains. A productive aquifer is also particularly advantageous to the growing urban centres within this geological region. Boreholes on the Trans-Amadi Industrial Estate at Port Harcourt yield 16,000 to 20,000 gallons per hour without difficulty, while an exceptional yield of over

---

[1] NEDECO, *River Studies and Recommendations on Improvement of Niger and Benue* (Amsterdam, 1959), Table 5.3.3–1, p. 488. (Hereafter referred to as *River Studies.*)

90,000 gallons per hour has been achieved.[1] The value of the aquifer diminishes only where the sands give way to clays, as at Opobo and Oron, and where they approach the coast, with the consequent risk of contamination through bringing in sea water. The productivity of the aquifer is also reduced on the margins of the formation, and any undue claims upon its resources (as has happened at Umuahia) can only bring about a depression of the water-table.

Offsetting their water-storage value, the Coastal Plains Sands contain no economic mineral deposits. The sediments of the Coastal Plains consist of unconsolidated coarse to medium-fine grained sands, with localised beds of fine sands and clayey shales. Occasional deposits of large water-worn pebbles and gravel beds of finer texture have been located during drilling and well-sinking. The base of the Coastal Plains Sands lies with a slight lack of conformity over almost all earlier formations from the Eze-Aku and Asu River Shales (Group vi), east of the Cross River (Fig. 4.1), to the rocks of the Bende-Ameki series (Group iii).

To the south-west, the margin of the Coastal Sands is difficult to distinguish since it merges almost imperceptibly (through the slightest change of gradient) with the alluvial deposits of the Niger Delta which in this area are largely reworked sands from the Coastal Plains.

(iii) BENDE-AMEKI GROUP

(d) *Lignite Formation.* These rocks consist of a series of sandstones, grits, shales, clays, with occasional beds of lignite or brown coal (the carboniferous remains of ancient forests). Topographically, the alternation of more resistant sandstones and grits with softer shales and clays has produced a series of parallel ridges and valleys which reflects in the distribution of population. The seams of lignite vary in number and thickness; two seams at Nnewi in Onitsha Province measure four feet and twelve feet six inches, rather thinner than those in the same formation west of the Niger River. Although a potential source of fuel, particularly as an alternative to firewood for household purposes, the Eastern Nigerian lignites are not worked at present.

Beds of plastic clay sixty feet thick or more provide promising material for the production of pottery, and pits have been sunk for the collection of trial samples.

(e) *Bende-Ameki Formation.* In the north-western section of this outcrop, within the Onitsha–Awka–Orlu triangle, the lithology of the formation

---

[1] Monkhouse, *Eastern Nigeria*, p. 9.

is dominated by the massive arenaceous (sandy) body of the Nanka Sands which is best developed in the Nanka-Nobi area, thinning to the south-east and disappearing at Okwelle.

The Nanka Sands consist of coarse and friable, cross-bedded white to yellow sandstones, with bands of red sandstone and sandy clays, dipping at a low angle (1° to 3°) to the south-west. They are underlain by the lower beds of the formation, comprising mudstones interbedded with shales and occasional thin limestones.

The outcrop of the Nanka Sands forms a plateau, falling away gradually to the west from the escarpment to the north and east, and tailing off to a narrow cuesta in the south-east where the sands are pinched out near Ok-welle. The topography over the rest of the outcrop consists of minor plateau and cuesta watersheds and parallel clayey depressions. Clearly these relief forms have attracted high densities of population along the ridges, with lighter settlement in the lowlands. Yet there is an acute lack of surface water on the water divides, due to the permeable sub-strata. There is only one major river, the Ideh-Mili, over the Nanka Sands, and some communities must seek water at distances of five miles or more from home, or at the bottom of the erosion gullies which scar the escarpment. Well below the surface, the Nanka Sands form a good aquifer, which is exploited by boreholes at several places, but the water-table is at depths of five hundred feet or more, and tapping this supply is expensive. Intensive settlement and over-farming of the Nanka Sands have induced serious sheet and gully erosion, particularly near the villages of Agulu and Nanka (Chapter 7).

In the south-eastern sector of the Bende-Ameki formation, north-west of Bende, the rocks are dominantly argillaceous (clayey) with interbedded sandstones and grits. Some of the clays are of value for pottery-making and a few pits are being worked for the ceramics industry. The dip of the strata is almost horizontal, and the countryside presents the picture of a low-lying and relatively deeply-dissected plateau, still well-forested and lightly settled due to heavy soils and the historic role of the Bende area during the slave trade and as part of the 'shatter-belt' between Ibo and Ibibio peoples.

(iv) IMO CLAY-SHALES GROUP

(f) *Imo Clay-Shales Formation.* This consists of impervious, unjointed clay-shales with occasional bands of clay ironstone and sandstone. There are also some large sand bodies interbedded along the centre of the out-crop which appear as strike ridges (north–south) in the otherwise flat and low-lying topography. The angle of dip in the northern part of the formation is barely 1° to the west, increasing to 3° in the southern part.

The impervious clay shales have induced a surface drainage pattern of dendretic rivers and streams such as the Imo and Mamu Rivers, which are particularly active in the rainy season. The poorly-drained heavy soils support few villages and most of the settlements are along the higher ground created by the intraformational sandstones. The clay beds are put to limited use in the fashioning of traditional pots, but there are few other economic uses for the rocks of this formation.

(v) PLATEAU AND ESCARPMENT GROUP

This largely Upper Cretaceous series of sandstones is responsible for one of the most prominent physiographic features in Eastern Nigeria, namely the Nsukka-Udi-Okigwi Plateau or Cuesta. These landforms are aligned in a north–south direction, but south of Okigwi trend eastwards, then southwards, to reach Arochuku via Ngusu in a subdued cuesta form (Fig. 4.1). This distinctive arrangement of uplands and associated features effectively separates the geological formations of the Cross River Basin to the east from those of the Imo Shales, the Bende-Ameki and Coastal Plains Sands to the west and south-west.

The rocks of the plateau and escarpment group comprise four concordant formations: the Upper Coal Measures, the False-Bedded Sandstones, the Lower Coal Measures and the Nkporo Shales (Fig. 4.3). All the beds dip to the west and south-west at a low angle to the north (1°), at greater angles towards the south (up to 10°). The line of strike swings in response to a series of post-Cretaceous warpings.

(g) *Upper Coal Measures Formation.* These rocks consist of fine-bedded white sandstones and siltstones, together with thin beds of coal, shales and limestones. The coal seams vary from a few inches to over five feet thick, but are jointed and inconstant which discourages their economic exploitation. The sandstones of the Upper Coal Measures are frequently indurated with iron oxides. This iron-enrichment or ferruginisation has been caused by poor drainage and ground water seepage, due to clay bands or shale partings in the sandstones.[1] It has resulted in a lateritic rock (laterite) which

[1] For a more scientific explanation of the processes of 'laterisation' and 'ferruginisation', see P. D. Jungerius, 'The Environmental Background of Land Use in Nigeria', *Tijdschrift van het Koninklijk Nederlandsch Aardrijkskundig Genootschap,* lxxxi (1964), pp. 422–4.

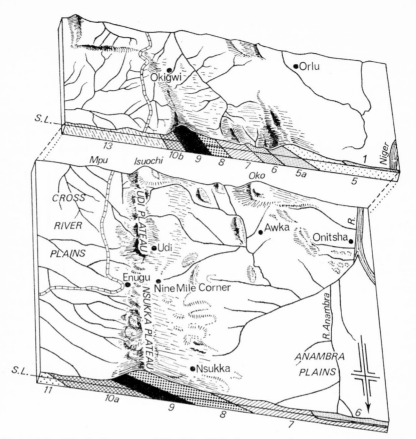

FIG. 4.3 *Block diagram of Nsukka-Udi plateau and escarpment zone (after A. T. Grove)*

1. Alluvium; 5. Bende-Ameki formation; 5a. Nanka Sands; 6. Imo Clay-Shales formation; 7. Upper Coal Measures formation; 8. False-bedded Sandstones formation; 9. Lower Coal Measures formation; 10a. Enugu Shales (Nkporo Shales formation); 10b. Awgu Sandstone (Nkporo Shales formation); 11. Awgu-Ndeaboh Shales formation; 13. Asu River Group.

is highly resistant to weathering and erosion. Acting as a capping stone, this ferruginised sandstone has induced some striking landforms in terms of an Upper Coal Measure Cuesta and associated buttes. The residual or outlier hills provided valuable hilltop defensive sites for villages in former centuries, while the lateritic rock made admirable building material for the stone-wall terracing of the steep slopes, for purposes of agriculture (Part I, Chapter 2, p. 46). The iron-enriched rocks of the plateau also form a reserve of potentially exploitable, very low-grade ore for smelting. Native artisans used these resources for agricultural implements and weapons in former days as is attested by the remains of smelting sites in scattered localities on the Nsukka-Okigwi Plateau.

(*h*) *False-bedded Sandstones Formation.* This formation comprises the principal outcrop of the Nsukka-Okigwi Plateau and consists of coarse-grained white and friable sandstones, with occasional thin bands of shales and grey clays. The sandstones are false-bedded on a large scale, although cross-bedding is a common feature of all Nigerian sandstones. It is these relatively resistant rocks, together with the Lower Coal Measures, which also form the main eastward-facing escarpment. Between Ukehe and Udi, however, severe gullying of these escarpment formations has taken place, probably due to stripping of the protective vegetational cover by shifting cultivators in former times.

Since the superficial strata are highly permeable there are very few rivers which traverse the plateau, although dry valleys indicate their former existence. Small dip-slope streams are fed by springs in favoured locations. However, the False-Bedded Sandstones form a capacious and high-yielding aquifer; urban centres such as Nsukka, and 'grassland towns' or super-villages such as Enugu-Ezike meet their water needs from this source. Westwards from the outcrop, drilling for artesian water has yielded free flows in excess of 6,000 gallons per hour from this aquifer. The elevated sandlands of the Nsukka Plateau with their free-draining soils have attracted a large rural population, as we have learned.

(*i*) *Lower Coal Measures Formation.* The lithology of the Lower Coal Measures consists of a rhythmic sequence of fine- to medium-grained white to grey sandstones, shaley sandstones, sandy shales and mudstones with – in the northern sector – coal seams or carbonaceous shales at various horizons.

The Lower Coal Measures make up the lower slopes of the scarp face running from the boundary with Northern Nigeria as far south as the Cross River near Arochuku (Fig. 4.1). East of Okigwi, however, the scarp and associated tablelands are much subdued, rising only 400 to 500 feet above sea level. All the same, this well-drained higher country proved attractive to settlement and provided an important communication route for the Aros and their allies.

The coal in this formation is economically important and represents the chief source of this fuel in all West Africa. Five potentially productive seams exist, numbered from one to five upwards. Due to diminishing demands and the rising costs of extraction, only the number three seam is worked at present, with an average thickness of four feet at Enugu. Unfortunately the coal is of inferior quality (sub-bituminous) and not suited for coking, i.e. as a source of fuel for smelting iron ore. Its traditional users have been the railways, but the challenge of diesel oil has had a depressant effect on the coalmining industry. The Oji River thermal power station mid-way between Enugu and Onitsha utilises Enugu coal, which is transported the intervening twenty-five miles by overhead cable buckets and by trucks. In addition to the coal, fire clay is worked near the old Obwetti Mine in Enugu. The clay is abundant and of good quality; it was formerly used for lining the fire-boxes of coal-burning locomotives.

(*j*) *Nkporo Shales Formation.* The area of outcrop of these rocks may be subdivided into four sections, essentially in a north–south direction. From the border between Northern and Eastern Nigeria, southwards beyond Enugu to near Ozalla, the lithology is of grey shales and mudstones with subordinate white sandstones. These beds are referred to as the Enugu Shales.

South of Ozalla to near Awgu, the shales are replaced by fine- to coarse-grained sandstones of various colours: the Awgu Sandstone member of the formation, which forms a part of the Enugu-Okigwi escarpment in this area.

South of Awgu, the sandstones die out and there is a return to shales, the Nkporo Shales, which turn eastwards following the base of the scarp as far as Afikpo. Here once again there is a strong development of arenaceous rocks in the Afikpo Sandstone. This resistant though strongly faulted and jointed series forms a further arcuate escarpment at Afikpo, paralleling that of the Lower Coal Measure and False-Bedded Sandstone scarp to the south-west. Finally, a little south of Afikpo, the sandstones are replaced by shales and this outcrop continues until overlain by the Coastal Plains Sands north east of Calabar.

(vi) CROSS RIVER PLAIN GROUP

Approximately twenty per cent of the area of Eastern Nigeria, coincidental with the upper drainage basin of the Cross River, is underlain by dominantly argillaceous, clayey rocks: impervious shales, mudstones, also porous limestones and micaceous sandstones (Fig. 4.1). These Lower Cretaceous sedimentaries are among the oldest in the region and are widely deformed through faulting and folding. This complex structure is essentially in response to the Abakaliki Uplift and the Afikpo Syncline. A further complication to the geology is provided by the volcanic dykes and sills, together with plugs and lava flows which appear in small outcrops scattered across the plain. Dyke and sill intrusions are fairly common in the Afikpo and Ugep areas and have been located as far west as Odomoke (near Nkalagu) and Lokpa Ukwu (south of Awgu). The biggest intrusive hilly area is that of the Workum or Leffin Hills on the Northern Nigerian border. Volcanic necks and plugs, also agglomerates and tuffs, are found in the vicinity of Abakaliki, while extrusive rocks in the form of basaltic lavas occur at Ikom and Obubra. It is interesting to note that, while these igneous rocks are considered to be of the Tertiary Age (probably Eocene-Oligocene), they are confined almost entirely to the strata of the Cross River Plain Group, and are not found in the Plateau and Escarpment rocks or other younger deposits.

The Cross River formations have relatively little resistance to erosion so that the general topography is that of a wide and slightly undulating plain, barely 200 feet above sea level, with a gentle slope towards the Cross River. The relief is broken only occasionally by hills formed in the more resistant rocks, the igneous intrusions and harder sandstones and limestones. While aquifers of useful size do not exist in the formation, surface water in the form of rivers, streams, shallow lakes and swamps is widespread.

The Cross River Plain Group is subdivided into three formations, aligned in a north-east to south-west direction as a result of the Okigwi-Abakaliki Anticline.

(*k*) *Awgu-Ndeaboh Shales Formation.* Outcropping only to the north of the line drawn from Okigwi to Abakaliki, these beds consist of blue-grey marine shales with subordinate shelly limestones and calcareous sandstones. Some of the limestone members near Enugu have recently been worked for road metal and aggregate, but the beds are not thick enough for large-scale operations. Along the northern tributaries of the Cross River which flow over the Awgu-Ndaboh Shales, e.g., the West Aboine, Nyaba and Asu Rivers, alluvial deposits derived from scarp retreat and gully erosion in their headwater areas are common.

(*l*) *Eze-Aku Shales Formation.* The rocks constituting this formation are a series of hard shales and silty mudstones, with economically significant limestones. A strong development of the latter at Nkalagu, some twenty-four miles east of Enugu, is being quarried for cement production by the Nigerian Cement Company (Nigercem). There are other deposits awaiting utilisation around Odomoke and at Igumale (Northern Nigeria), also further east at Bansara and Ogoja. Some of the shales may also be usable for the manufacture of bricks and crude pottery. These rocks also outcrop to the south of the Cross River where limestones and shales near the large indigenous town of Ugep await commercial development.

(*m*) *Asu River Group.* The lowest unit in the Cross River Plain series and the oldest sedimentary rocks to outcrop in the region are those of the Asu River Group. Marine shales with interspersed sandstones and limestones have accumulated to a maximum depth probably in excess of 6,000 feet. The limestones of this group have not been worked although the shale deposits near Awgu were at one time used for brickmaking. Also near Awgu are workable deposits of galena (lead ore), while galena and sphalerite (zinc ore) near Abakaliki are periodically exploited when world prices are favourable.

(vii) BASEMENT COMPLEX

(*n*) *Basement Complex Formation.* Outcrops of metamorphic and igneous rocks – schists, gneisses, and granites – so common elsewhere in West Africa and throughout the continent, appear only in minor areas of Eastern Nigeria, covering at best 15 per cent of the region. Yet they have created distinctive landscapes and have had a marked influence on human settlement patterns and utilisation of the areas concerned.

The age of the Basement Complex rocks has traditionally been described as Pre-Cambrian, i.e. dating from perhaps 500 million years ago, while certain of the schists have been considered Archaean in age (among the oldest in the world), dating back nearly 3 billion years. Recent geological investigations utilising radioactive dating techniques[1] have however suggested that two of the major series of rocks, the Older and Younger Granites, may be much younger, dating from the Lower Palaeozoic and mid-Jurassic Ages respectively. Further studies will be necessary to unravel the full geological history of these intriguing formations. The probability is that much of the Basement Complex will still prove to be Pre-Cambrian, while lesser areas may be shown to comprise rocks of a later age.

It is certain that, whatever their ultimate age, the rocks are all ancient when compared with the region's sedimentary formations, and have been

---

[1] See R. A. Reyment, *Aspects of the Geology of Nigeria* (Ibadan, 1965).

subjected over aeons of time to successive mountain-building orogenies and subsequent cycles of erosion. The present elevations of the mountains, plateaux and hills in the Obudu and Oban Uplands were probably attained during the late Tertiary period as a result of widespread continental movements at that time.

The contact between Basement Complex rocks and surrounding sedimentaries is not clear-cut (at least to the untutored eye), with the exception of the area north of Calabar where, on the south-west margin of the Oban Plateau, the succession of outcrops along the road from Calabar to Oban is both rapid and not difficult to identify (Fig. 4.1). Apart from local quarrying for road metalling and foundation material, these rocks have not proved economically productive.

5 Landforms, Relief and Associated Drainage Features

INTRODUCTION

THE principal characteristics of the land surface of Eastern Nigeria are the limited areas of hills and mountains and the dominance of level to moderately rolling plains. The Niger Delta, the broad Coastal Plain and the Cross River Basin together account for almost seventy per cent of the region. Elevations over the greater part of the lowlands seldom exceed 400 feet above sea level (Fig. 5.1). In few other parts of West Africa are there coastal lowlands of comparable dimensions.

The absence of major relief obstacles, also rapids and waterfalls, has meant that movements overland or by water have always been relatively simple throughout this part of West Africa, both for indigenous peoples in prehistoric and historic times, also for alien groups in recent centuries. Indeed:

> The Coastal Plain of Nigeria has served as one of the most important areas of penetration into Africa ever since contact with Europe by sea was established. . . . The level landscape and the well-drained soils of the Coastal Plain have encouraged the movement and settlement of population which has turned Southern Nigeria into the most densely settled area of tropical Africa and provided it with the densest road network on the continent.[1]

As indicated in Chapter 4 (Geology), the land of Eastern Nigeria ascends slowly and gradually from the Niger Delta in the south-west (0 feet) to the mountains of the Obudu Plateau in the north east (maximum elevation 6350 feet). Approximately at right angles to this line of ascent, the pattern is firmly interrupted by the pronounced physiographic features of the Plateau and Escarpment zone, also the Oban hills, which together form a north-west to south-east axis of markedly diverse relief and topography.

Five broad physiographic regions may be recognised in Eastern Nigeria.
   (i) The Deltas and River Plains
  (ii) The Coastal Plain.

[1] Govt. of Eastern Nigeria, *Water Supply*, Section III, pp. 1–2.

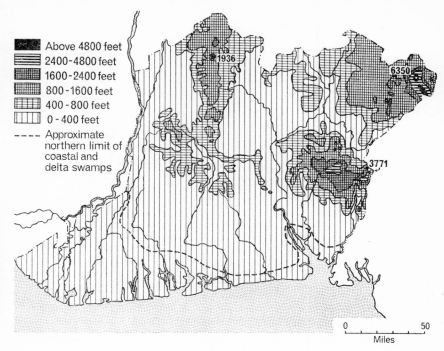

Legend:
- Above 4800 feet
- 2400-4800 feet
- 1600-2400 feet
- 800-1600 feet
- 400-800 feet
- 0-400 feet
- ---- Approximate northern limit of coastal and delta swamps

1936
6350
3771

0 _____ 50
Miles

FIG. 5.1 *Relief*

(iii) The Zone of Plateaux and Escarpments with associated lowlands.
(iv) The Cross River Basin.
 (v) The Eastern Uplands.

(i) THE DELTAS AND RIVER PLAINS

(*a*) *The Niger Delta.* As the Greek name implies, the main feature of a classical delta is a triangle formed by the distributaries of a river as it approaches the sea. The vast Niger Delta, like that of the Nile (which it challenges in size) fully warrants its appellation (Fig. 5.2).

The apex of the Delta, near the village of Aboh, is about 150 miles inland from the Gulf of Guinea. South of Aboh, there is a remarkable change in the behaviour of the Niger. Bifurcations, ramifications and meanderings of the river commence until its waters complete their 2,550-mile-long journey from the Fouta Djallon Plateau (Republic of Guinea) to the sea, discharging through many outlets. The Delta continues to grow seaward

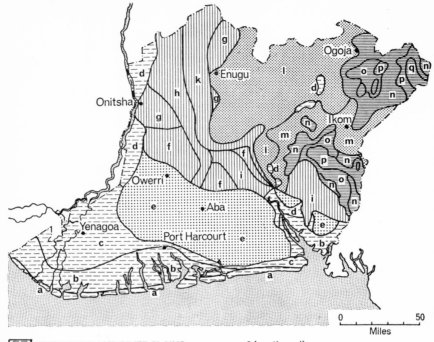

1 [legend] THE DELTAS AND RIVER PLAINS
  a Sandy beaches and ridges
  b Tidal creeks and mangrove swamps
  c Sandy delta plains and freshwater
    swamps
  d River flood plains
2 [legend] THE COASTAL PLAIN
  e Flat coastal plain
3 [legend] ZONE OF PLATEAUS AND ESCARP-
  MENTS, WITH ASSOCIATED LOWLANDS
  f Rolling dissected lowlands
  g Strongly dissected plateau, medium slope
  h Undulating lowlands

3 (continued)
  i Incised diversified lowlands
  k Plateau extending above 1200 feet,
    gentle dip slope
4 [legend] CROSS RIVER BASIN
  l Slightly incised plains
  m Rolling plains and dissected hills,
    above 400 feet
5 [legend] EASTERN UPLANDS
  n Rolling plateau above 800 feet
  o Dissected plateau above 1200 feet
  p Strongly dissected plateau above
    1600 feet
  q Rugged mountains above 3000 ft.

FIG. 5.2 *Landforms (modified from Y. Karmon)*

since the major part of the transported sediments is carried as far as the coast.

The Delta may be divided into three sections:

(1) An area of sandy beaches and ridges along the coast.

(2) An area of tidal creeks and mangrove swamps.

(3) A seldom-flooded, sandy plain in the upper reaches.

(1) The line of sandy coastal ridges varies in width from less than one hundred feet up to ten miles; the beaches are from two to five feet above high-water level. While subject to slight erosion on the landward side, their size and shape is controlled largely by the eroding and accreding forces of the littoral drift. The prevailing direction of wind and swell in the Gulf of Guinea is south-south-west to south-west, and the surf hits the coast at right angles near the Nun and Sengana entrances to the Delta.[1] On both sides of this point, the angle between the wave crests and the coastline causes a littoral drift of sediment. Due to a slight skewness to the distribution of the Delta (resulting from preferences given by the Niger to certain outlets over geologically short periods), a rather larger amount of the river's sand load has been deposited over the western part of the Delta. The sand beaches are frequently interrupted by tidal creeks, although they usually continue as underwater sand bars across the mouths of the creeks.

> Owing to their elevations and the storage of rain water within the sands, they are suitable for human habitation. In the colonial history of the Region the beach settlements have played a very important part in providing the first bases for foreign penetration and trading or government stations, and some of them have assumed the dimensions of towns.[2]

Examples of such coastal communities are Brass, Bonny and old Opopo, the historic slave-trade ports of the Bight of Biafra.

(2) Behind the coastal region, the tidal creeks and mangrove swamps have an average width of fifteen miles, and are subject to constant inundation by saline waters. The semi-diurnal tide enters the swamp area in at least ten different places and, with an average rise of up to seven feet in the eastern part of the Delta, it is estimated that about 150×10^9 cubic feet of water flow in and out during every cycle of the tide.[3]

The anastomosing complex of creeks, with an average depth of thirty feet, are navigable by small vessels although they twist and turn neverendingly around the large number of mangrove-covered, muddy islands, and are shallower at points where the tidal currents converge. The area is in general sparsely settled (Fig. 2.2); Ijaw fishing communities are found along the banks of creeks where sandy accumulations have created slightly higher ground. At a few favoured sites, however – e.g. Buguma, Abonnema

[1] NEDECO, *River Studies*, Section III.2, p. 268.
[2] Govt. of Eastern Nigeria, *Water Supply*, Section III, p. 9.
[3] NEDECO, *River Studies*, p. 263.

and Oloibiri – towns of considerable size have emerged. It is overwhelmingly a watery environment, for to the ever-present creeks, bayous and swamps of the Delta must be added an annual rainfall of between 100 inches and 140 inches and no appreciable dry season.

(3) The sandy area in the upper reaches of the Delta extends downstream for about seventy-five miles from its apex. Physiographically, it is a continuation of the flood-plain found along the lower Niger from Aboh northwards to the boundary with Northern Nigeria and beyond. The braided nature of the river gives way south of Aboh to an undivided channel with marked meandering until, just below Samabri, the Niger divides into the Forcados and Nun Rivers (the latter being Eastern Nigeria's principal distributary of the Niger), and each in turn bifurcates again and again into meandering river branches until they enter the tidal mangrove swamps. The innumerable waterways have thrown up natural sand levees in the upper reaches of the Delta, which protect the intervening lowlands from flooding. However, ground water is so near the surface across most of the area that fresh water swamps and forests are widespread. Settlement is again light, being confined to higher and drier ground, and intercourse is largely by water. There is regular river traffic between Onitsha and Northern Delta towns such as Odi, Kaiama, Sabagreia and Amassoma.

(b) *The Niger Flood Plain.* Southwards from the intersection of the boundary between Northern and Eastern Nigeria and the Niger River, the Niger forms the western limit of a twenty-mile-wide interfluvial plain of alluvium, bounded by the Anambra river and its sediments on the east. The Niger-Anambra Plain is a confusion of old alluvial terraces, meander scars, ox-bow lakes and natural levees. The joint flood plain eventually tapers to a point at Onitsha, where the river valley narrows and is contained within a probable fault in the Bende-Ameki Sandstones (Nanka Sands). At Onitsha, which is the largest town on the Niger, the annual discharge of the river reaches its maximum, averaging $7,000 \times 10^9$ cubic feet, with a distinct peak discharge in October of about 800,000 cusecs.[1]

After negotiating the rocky restriction, the river again flows through a broad flood plain of its own alluvium deposits until reaching the Delta. Additional potential tributaries from the east are prevented from reaching the Niger by the parallel-flowing Orashi River, which, rising in the rolling coastal plains south-east of Onitsha, shares the Niger flood plain for a

[1] Ibid., Section III.2, p. 247.

considerable distance before finally being absorbed into the Delta drainage complex. Extensive areas of the plain are inundated when the river is in flood, except for certain alluvial terraces which stand above the high-water level of the swollen river (a thirty-foot annual rise is not unusual). Natural sand levees paralleling the river provide higher ground for village settlements but due to perennial flooding intercommunication is limited to dugout canoes (sometimes fitted today with an outboard motor) and there are no roads within the flood plain proper. A low density of population prevails throughout the area. (Fig. 2.2).

(c) *The Cross River Delta.* This feature is, by comparison with the Niger Delta, a far less significant physiographic phenomenon. Commencing at a point on the river east of Uyo, the small delta is covered with mangrove swamp forests, being already within range of the tide. The drowned Cross River estuary is joined by the mouth of the Calabar River; other smaller rivers flowing southwards from the Oban Plateau, e.g. the Great Kwa and Akpa Yafe, also join the delta zone of the Cross River and discharge their waters into the broad conspicuous embayment.

(d) *The Cross Flood Plain.* The flood plain along the lower Cross River is limited owing to the fact that it is hemmed in on the east by the crystalline mass of the Oban hills and on the west by the tail-end features of the tapering Plateau and Escarpment zone. Only when it enters the Coastal Plain region, south-east of Arochuku, is there a chance for a flood plain to form. Levees and alluvial terraces are present, but the upper reaches of the irregular delta are soon met. Settlement is sparse, being limited to fishing communities in remote localities.

(ii) THE COASTAL PLAIN

Over this extensive plain, relief is but a few hundred feet while local relief may be measured in tens of feet; physiographic differences are thus negligible.

Only north of a line joining Onitsha, Owerri and Ikot Ekpene does the flat, featureless nature of the landscape yield to rolling and dissected lowlands. The modest change in the appearance of the terrain results from minor scarping of the Nanka sandstones in the Bende-Ameki series along the line of contact with the Imo Shales. Gentle dip slopes have induced dissection by the few surface streams flowing across the Nanka Sands. These physiographic and lithologic factors warrant the inclusion of the resultant rolling plains within the zone of Plateaux and Escarpments with associated lowlands. They are, in fact, marginal and transitional features. It is in this region of dissected lowlands and associated interfluves that, as we have learned, the highest densities of non-urban population in Eastern Nigeria are experienced, in excess of 1,500 persons per square mile in parts of Orlu and Okigwi Divisions.

Few rivers and streams originate within the flat coastal plain, due not merely to insignificant relief but also to small run-off as a result of highly permeable sub-strata. There are only four main rivers – the Otamiri, Imo, Aba and Kwa Ibo – and one of these (the Imo) has its source within the plateaux and lowlands to the north. The longest and most interesting river, in point of its unusual and irregular course, is in fact the Imo. Rising in the southern sector of the clay-shales belt of the same name, the Imo River at first flows south-eastwards towards the Cross; then it turns abruptly south-west and follows an almost perfectly straight course for over fifty miles towards Port Harcourt. But after its confluence with the Otamiri River, it again turns at right angles, to debouche into the Bight of Biafra, thirty-five miles east of the Bonny River estuary.

The behaviour of the Imo River illustrates a clear case of river capture. The upper section, on the Imo Shales, is a strike river flowing through a broad valley lying north-east of the Nanka Sands escarpment and south-west of the Okigwi Cuesta. This valley continues to the Cross River, widening as it goes, but is occupied in its lower reaches by a small misfit stream, the Enyong Creek, which represents the atrophied remnant of the former Imo River. North-west of Umuahia a dip-slope river, cutting back at unusual speed across the 500-foot escarpment, penetrated the Imo valley and led its waters off southwards into the Coastal Plain. According to Pugh, the capturing stream followed a fault line, which accounts for its extraordinarily straight course in the middle section of the present Imo River.

> Arguments in favour of such a hypothesis [are] the remarkable linear pattern, the absence of side streams (suggesting very rapid headward erosion) and the parallelism between this stretch and the known lines of weakness in the Cameroons to the east. Traditional Ibo legends of a sudden appearance of the stream may throw some light on the origin of the river.[1]

The incised nature of the river, together with its frequent minor meanderings, are well illustrated on a large-scale map and appear to support Pugh's

[1] J. C. Pugh, 'River Captures in Nigeria', *The Nigerian Geographical Journal*, iv (Dec. 1961), p. 43.

thesis of its origin. However, the reason for the second abrupt change near Port Harcourt has yet to be satisfactorily explained. It is possible that, at its southern extremity, the fault valley has been blocked by reworked alluvial deposits of the Niger, of which the lower Coastal Plains sands are largely comprised; this would result in deflection of the Imo River eastwards.

One other interesting hydrographic feature in the flat Coastal Plain is Oguta Lake, in the extreme north-west of the area. This lean 'finger lake', which exceeds ten miles in length, has been formed by the damming of the lower Nyaba River with alluvium deposited in the Niger Flood Plain, at the point of juncture of the Nyaba with the Orashi River. No alternative discharge route has developed on the surface, although sub-surface effluence is believed to be strong.

(iii) THE ZONE OF PLATEAUX AND ESCARPMENTS WITH ASSOCIATED LOWLANDS

This area of complex geomorphological structures is made up of a sequence of plateaux or cuestas, clearly-defined escarpments and included lowlands (Fig. 4.3).

The core features of the zone are the Nsukka-Udi Plateau which continues southwards in subdued form as the Okigwi Cuesta, then turns eastwards to Ngusu (near Afikpo) and finally southwards to Arochuku; also the pronounced escarpment associated with the Plateau, which can be traced from the boundary of the Northern Provinces all the way to the Cross River Flood Plain. The main Nsukka-Udi Plateau and Escarpment is separated by a belt of undulating lowlands from a secondary plateau and scarp region to the west, beyond which lie the flood plains of the Anambra and Niger Rivers.

Description of these landforms and the resultant landscapes is best achieved by dividing the zone into northern and southern sections, with Okigwi as the point of separation.

(a) *The Northern Section.* The False-Bedded Sandstone plateau is at its broadest in the north, around Nsukka; southwards, via the Udi and Awgu sub-regions of the plateau, this impressive physiographic feature narrows until, near Okigwi, it is more properly termed a cuesta (Fig. 4.3). The escarpment on the eastern side of the plateau, capped for much of its length by a resistant white sandstone of the Upper Coal Measure series, varies in height from 500 to 900 feet. The scarp face is seldom sharp, except for the almost perpendicular slopes of the massive capping rock; below the False-Bedded Sandstones, the less resistant Lower Coal Measures induce concave slopes with scree or erosional debris at their base.

Similarly, the line of the escarpment is by no means a regular one, either in profile or in orientation. Subsequent streams have contributed to undercutting of the capping sandstone and its falling away in joint blocks. Spurs, re-entrants and interconnecting saddles or cols (the results of differential erosion) add to the irregularities of the scarp, inducing a sinuous course and an uneven crest line. In places, isolated knolls or buttes have been left in front of the scarp face, e.g. 'Juju Hill', Enugu.

Headward-cutting obsequent streams fed by springs have carved out great scallop-shaped gullies in the False-Bedded Sandstones and Lower Coal Measures of the escarpment. This secular erosion has been accelerated by human clearance of the former rainforest from the plateau and escarpment. With the sudden change of slope, great accumulations of sand choke the lower course of the scarp streams. Rivers such as the Aboine, Awra, Idodo, Ekulu and Nyaba have, as a result, highly braided channels; in the dry season they are largely sand rivers, with little surface movement of water.

N. P. Iloeje has observed that these rivers are of great interest to physiographers in as much as they spatially demonstrate over a distance of less than ten miles all stages in the life cycle of more normal rivers.[1] Thus the headwater gullied sections have youthful, V-shaped valleys and ravines; the middle sections (still in the scarp zone) have a more mature profile and some meandering, while the sand-filled lower sections at the base of the escarpment, with flat profiles and broad, terraced flood plains, symbolise old age. Of considerable interest too is the series of small lakes found along the margins of the alluvium-filled, obsequent river plains. Many of these result from the damming-up of small side valleys on the undulating, often dissected surface of the upper Cross River plain immediately to the east of the escarpment, underlain by Enugu Shales (the origin of these lakes is thus similar to Lake Oguta in the Coastal Plain region). Limited catchment areas feed the naturally-created reservoirs which provide a useful year-round source of surface water for the local population. Others of these lakes within the flood plain proper are no more than detached sections of the main rivers, from which they have been cut off by alluvial deposition. Such pools continue to be seasonally charged with flood water.

A modification in the interplay of physiographic and geological factors

[1] N. P. Iloeje, 'The Structure and Relief of the Nsukka-Okigwi Cuesta', *The Nigerian Geographical Journal*, iv (August 1961), p. 23.

along the Nsukka-Okigwi escarpment occurs in the vicinity of Awgu. Here, for a distance of twenty miles (between Ogugu and Lokpanta), Awgu Sandstone takes the place of the white sandstone as the capping rock. This forms a steep and even more pronounced scarp face than that further north, for obsequent streams have thus far failed to make much impression on these erosion-defying formations.

In echelon with the Awgu Sandstone escarpment, a subdued False-Bedded Sandstone scarp lies some three miles behind it to the west, dying away gradually to the north and south. The intervening basin of Lower Coal Measures and Nkporo Shales has been penetrated by the headwaters of the consequent dip-slope Mamu and Oji Rivers to create a hilly, deeply-dissected landscape; it is this area which has seen the development of some of the most remarkable terraced agricultural landscapes in Eastern Nigeria (Part III, Chapter 10).

To the west of the escarpment, the northern sector of the Plateau and Escarpment Zone may be further subdivided by an east–west line passing through Enugu and Nine-Mile Corner (Fig. 4.3). This line coincides with the point where the headwaters of the Ekulu and Iva Rivers, draining to the Cross River via the West Aboine River to the east, almost reach the headstreams of the Mwuyi River, a tributary of the Mamu, draining to the Niger on the west. North of the line lies the Nsukka Plateau, an ever-widening upland of rolling hills and broad valleys. To the south lies the narrower Udi Plateau and Awgu-Okigwi Cuesta.

The Nsukka Plateau is the broadest as well as the highest section of the entire complex of landforms in the north-western part of Eastern Nigeria. The average width of the Nsukka Plateau is thirty miles, while the crest averages 1,500 feet above sea level. Further superimposed upon it are Upper Coal Measure hills capped with ferruginised sandstone which reach heights in excess of 1,700 feet above sea level. (The highest elevation is 1,936 feet, one mile south-west of Ukehe).

The top of the plateau slopes gently to the west and comprises a number of broad sandstone ridges divided by east–west orientated dry valleys containing superficial sandy deposits. The headwaters of only two consequent rivers, the Adada and the Ijali, cross the Nsukka Plateau. Settlement on the plateau is dense, averaging over 600 people per square mile in Nsukka Division, and reaching over 1,300 persons per square mile around Enugu-Ezike; lack of surface water creates serious problems for the village populations in the dry season and has necessitated the sinking of deep wells and boreholes, and construction of water-storage facilities and standpipes

The most striking physiographic features are unquestionably the large number of small yet pronounced hills on the Nsukka Plateau, broadly within a ten-mile zone bounded on the east by the main road leading from Nine-Mile Corner to Northern Nigeria. Towards the west, these isolated hills or buttes combine into mesas and eventually form a minor cuesta which marks the start of a gradual descent westwards to the undulating Imo Shale Lowlands and the Anambra Valley. Like the more pronounced scarp face to the east, the Upper Coal Measure Cuesta has an irregular, serrated edge. But contrary to the Enugu escarpment, where obsequent streams have played a dominant role, consequent streams have modelled the Upper Coal Measure scarp and its associated residual hills or outliers. The even-crested ridges of the escarpment and the many isolated hills scattered in front of the scarp face are remnants of the Upper Coal Measures which once completely covered the watershed between the consequent rivers. These rivers, cutting through the impervious, resistant layer of ferruginised sandstones, eventually reached the highly permeable False-Bedded Sandstones and disappeared underground, leaving dry valleys. The valleys have continued to widen through undercutting and recession of the capping stone. They show no trace of the former river bed and the present water-table is situated far below the bottom of the valley.

The profile of the escarpment slope in the dry valleys is characterised by two distinct breaks in the slope. Just beneath the flattish surface of the cuesta, an almost perpendicular slope occurs. This changes to a steep talus slope, and another pronounced break at the foot of the scarp is observable, leading to the smooth slope of the pediment. This sequence of slopes is commonly found in landforms within arid areas; they are unusual within the humid tropics. As Jungerius has observed,[1] they result from the distinctive lithological sequence in the area. Elsewhere in West Africa, the resistant formations in cuestas and plateaux are permeable while the underlying weaker rocks are relatively impermeable. Here on the Nsukka Plateau the reverse obtains, leading to landforms not usually produced in tropical rainy climates.

The hilly outliers assume three forms. Some are flat-topped and tabular in shape (e.g. near Obimo, south-west of Nsukka). The protective layer is still present on the summits of these hills and the slope are concave, retreating laterally as the resistant capping stone is undermined. As indicated in earlier chapters, these flat-topped outlier hills provided

---

[1] P. Jungerius, 'The Upper Coal Measures Cuesta in Eastern Nigeria', *Zeitschrift fur Geomorphologie*, v (1964), pp. 167–76.

excellent natural sites for fortified hilltop residences, particularly for Igala settlers in the eighteenth century, and are still occupied by villages in some locations today.

With the complete removal of the lateritic layer, the hills assume a conical shape (e.g. near Leja, south of Nsukka). They begin to wear down rapidly, although the concavity of the slope is still maintained. With continued weathering and erosion, and a slowing down of lateral retreat, the hills are lowered and become rounded at the summits (e.g. near Enugu-Ezike, north of Nsukka). The upper slopes are now convex while the lower slopes remain concave. Eventually the hills are reduced to mere swells on the landscape although their former position is invariably marked by a concentration of ferruginous boulders and red gravelly soils. Successive stages in the evolution of these landforms, according to Iloeje, are shown in Fig. 5.3.

STAGE A   STAGE B   STAGE C
YOUTHFUL   MATURITY   EARLY OLD AGE

FIG. 5.3   *Suggested stages in profile evolution of outlier hills on the Nsukka plateau (after Iloeje)*

South of the Enugu–Nine Mile Corner dividing line lies the Udi Plateau and the Awgu-Okigwi Cuesta. Around Udi, the Plateau is considerably narrower than around Nsukka. From a maximum width of perhaps twenty miles at Udi, it tapers southwards to a cuesta form near Awgu. The height of the plateau edge averages 1,100 feet above sea level, with occasional hills reaching over 1,200 feet, Between Udi and Awgu, the plateau drops eastwards in two steps, with a rolling lowland area between the two scarps. The dip slope is dissected by the headwaters of the Oji and Mamu rivers, which originate almost at the edge of the main scarp, and by those of the Imo River, north west of Okigwi. Sloping gradually westwards and decreasing to an elevation of 500 feet, the Upper Coal Measures of the Udi Plateau merge finally into the lowlands of outcropping shales of the Imo River series.

(b) *The Southern Section.* The southern sector of the Plateau and Escarpment Zone extends from Okigwi eastwards then southwards to Arochuku.

It is rather lower and narrower than the northern sector. The scarp crest averages only 500 feet above sea level, although a few hills near Okigwi and Ngusu reach 1,000 feet; in width the hilly area is scarcely more than ten miles across, except near Afikpo (on the axis of the Afikpo syncline) where it broadens to twenty-five miles.

The False-Bedded Sandstone Cuesta carries the main trunk road in a sweeping arc from Okigwi to Ovim and Ngusu, thence to Ebem and Arochuku; in its alignment, the modern road follows the former route from the north leading to the 'Long Juju' of the Aros. The Lower Coal Measure sandstones of the scarp are softer than those further north, producing gentle slopes. Obsequent streams drain to the Asu and Cross Rivers while dip-slope streams feed the Imo and Enyong rivers. Between Arochuku and the Cross River, the tail-end of this pronounced Eastern Nigerian group of landforms may correctly be termed a hogback ridge.

The Imo Shale Lowlands, to the west of the main Plateaux and Escarpments, are relatively featureless by comparison. The level to undulating terrain of the soft clay shales is interrupted only by low strike ridges of intraformational sandstones (e.g. east of Awka) and small ridges of terrace gravels along the many incised water courses. The impermeable soils induce a high rate of run-off and perennial streams creating a dendretic drainage pattern are common. Due to extensive horizontal tracts in the Imo Shale Lowlands, swamps develop in the rainy season, inhibiting agriculture.

The westernmost scarps and related plateaux or cuestas result from the interbedding of resistant and non-resistant rocks of the Bende-Ameki series of sandstones, shales and clays. Parallel ridges and valleys trend in a generally north–south direction, providing ideal sites for defensive settlements in former years. The principal feature is the Nanka Sandstone Plateau which attains an altitude of over 1,000 feet and is eight miles wide at its broadest extent, tailing off to a narrow cuesta in the south-east (Fig. 5.2). The sandstone escarpment south of Awka has been severely dissected by obsequent, spring-fed streams whose headwater areas have awesome erosion gullies and ravines, particularly in the vicinity of Agulu, Nanka and Oko (Chapter 7).

Only one main consequent stream, the Ideh-Mili, crosses the Nanka Sands Plateau to join the Niger River south of Onitsha. Mass wasting and erosion have dammed the headquarters of this river to create the Agulu Lake, a useful natural reservoir for the dense population in its vicinity. The south-western margin of the plateau is deeply dissected by tributaries of the Orashi River and falls away sharply to the Niger flood plain.

To the south-east, from Okwelle to the Imo River, a constriction of the cuesta landforms takes place; but east again, beyond the river, the uplands broaden to a dissected plateau around Bende, separated by only a few miles from the Okigwi-Ngusu-Arochuku Cuesta and related features.

(iv) THE CROSS RIVER BASIN

More than one-third of Eastern Nigeria is drained by the Cross River and its numerous tributaries (Fig. 5.1). Rising in the Cameroon Mountains, the Cross flows for about one-half of its 250-mile course in a north-westerly direction; only after entering the Eastern Provinces and receiving an important northern tributary, the Okpauka (Anyim) river, does it turn south-west and finally south-east to enter the Bight of Biafra south of Oron and Calabar.

Due to the impervious nature of the ground and the high rate of run-off, streams are numerous throughout the Cross Basin. They form a closely dendretic, mature drainage pattern and are possibly superimposed. Those originating in the river plain itself are largely seasonal; only major streams rising on the Nsukka-Okigwi escarpment to the west or the crystalline uplands to the east and south-east carry water the year round.

The Cross is markedly off-centre in relation to its watershed, so that the principal perennial tributaries – all incised to a depth of twenty feet or more – come from the north: the Asu and West Aboine, East Aboine, Okpautu and Aya. Only short streams reach the Cross from the Oban Plateau. As observed earlier, the escarpment rivers, rising between Oboli and Udi, are flanked by strings of lakes from a distance of about ten to fifteen miles downstream from the foot of the scarp, created by the damming of side streams by alluvium in the aggraded stream beds. These lakes are often half a mile in length, but fluctuate considerably from the wet to dry seasons.

Topographically, the Cross River Basin comprises a plain with an average elevation of 200 to 300 feet, sloping gradually to the south-east. The highest areas in the Basin lie to the north-east, on either side of the Okpautu river system. The undulating to gently rolling surface is broken by occasional sharp ridges resulting from sandbodies, particularly in the western section of the Basin, and isolated igneous hills, increasing in number towards the north-east. The largest intrusive hilly area occurs near the boundary with the Northern Provinces, the Workum or Leffin Hills. Better known is the volcanic intrusion near Abakaliki, which gives rise to a prominent hill on the north side of town (a natural site for the town's water-storage system). Few of these scattered hills were large enough to support fortified villages in earlier times; this lack of defensible sites contributed to the depopulation of the Cross River area during the days of slavery.

The lower course of the Cross River, south of the Basin proper, is aggraded and bordered by lakes and swamps. Lake Ukwa, north-east of Arochuku, is the largest in Eastern Nigeria, having a surface area of approximately eight square miles. The Enyong River or Creek is the last long right-bank tributary of the Cross, and drains the lower reaches of the former Imo River.

(v) THE EASTERN UPLANDS

An elevated complex of crystalline hills, mountains and plateaux rises out of the plains in the eastern and north-eastern sections of Eastern Nigeria, and stands in stark contrast to the flattish terrain of the Cross River Basin and Coastal Lowlands.

Accordance or near-accordance of summits along several erosional surfaces indicates that today's landforms represent merely the present stage in the geomorphological evolution of formerly extensive plateau massifs which have probably undergone several cycles of erosion, peneplanation and re-elevation. The two areas of contemporary uplands are separated by the broad trough of the upper Cross River (Fig. 5.2).

On the north, largely in Obudu Division of Ogoja Province, a planated margin of rolling and dissected plateau country from 800 feet to 1,600 feet above sea level divides the Cross River Basin from the more rugged and strongly carved surfaces of the Obudu Plateau, from 1,600 feet to over 5,000 feet above sea level. Individual peaks attain elevations of over 6,000 feet, e.g. Kolo Ishi (6,350 feet), Kolo Koshun (6,250 feet) and Bebi Mountain (6,100 feet). The landscapes within the Plateau proper are truly impressive, consisting of several parallel-trending ranges of mountains separated by deep, youthful valleys.

The elongated hill and mountain ranges throughout the north-eastern uplands trend in a generally north-north-east to south-south-west direction, with the principal river valleys, broad and flat-bottomed in the planated marginal zone, following suit in a characteristic ridge-and-valley fashion. Deviation from this alignment occurs, as in the case of the Sankwala Mountains and the main mass of the Obudu Plateau, topped by Kolo Ishi and Kolo Koshun to the south, where an east–west direction is assumed. These deviations may be caused by localised intensive earth disturbances due to faulting, granitic intrusions, or a combination of both. East of Obudu township, the hill ranges swing to the north-west; while north of the town, in the Tiv or Munshi country, they gradually disappear, leaving only occasional low granitic *inselbergs* to mark their former existence.

The drainage pattern is distinguished by mature subsequent or strike streams which follow the flat valleys separating mountain ranges; these rivers have developed along lines of weakness afforded by foliation trends in the highly metamorphosed rocks and, together with their obsequent and resequent tributaries, form a trellis type of drainage, e.g. the Sankwala and Aiya Rivers, which drain north eastwards to the Katsina Ala in Northern Nigeria, thence to the Benue River. Elsewhere, when rocks are more homogeneous and massive, as in granitic terrain, a dendretic drainage pattern has developed; e.g. the headwaters of the Aya River, a northern tributary of the Cross. The rugged landscapes and generally steep slopes of the Obudu Plateau have made agriculture and communications difficult and have thus inhibited human settlement (Fig. 2.2).

South of the Cross River, the Oban Plateau is a somewhat subdued reflection of the Obudu Uplands. A series of flat-topped plateaux reaches elevations of over 3,000 feet, with the highest point at 3,771 feet, close to the Cameroon border. These highlands are again separated by very steep river valleys, which drain both north to the Cross, and southwards, as in the case of the Calabar and Greater Kwa rivers, to the Cross River estuary. Since, on the southern side of the Oban Hills, igneous rocks lie high above the Cretaceous and Tertiary sedimentary formations towards the coast, the mature rivers fall off the plateau in cascades and waterfalls, offering potential hydro-electric power sites.

The Oban Plateau has steep flanks to the north and south, but drops more gradually westwards through the planated margins to the middle and lower sections of the Cross River. Although the relief is generally lower than the Obudu Plateau, the Oban topography is no less difficult to traverse due to the deeply incised streams; human settlement is even sparser here than in Ogoja Province. Indeed, as we have observed, the Oban Uplands and parts of the Niger Delta represent the only sizeable remaining areas in Eastern Nigeria where the frontiers of human settlement have yet to penetrate.

D

6 Soils[1]

INTRODUCTION

A knowledge of the characteristics and location of the major soil groups in Eastern Nigeria is important to a fuller understanding of the present and former patterns of agricultural land use and rural settlement within the region; such a knowledge may also aid in evaluating the prospects for improvements in farming and alleviation of the problems of over-population through resettlement and other development schemes.

The science of pedology has long been dominated by concepts of soil genesis developed by Russian and American soil scientists in the late nineteenth and early twentieth centuries, based upon field studies within the intermediate latitudes of the Old and New Worlds. Considerable emphasis has been given to factors of climate (particularly precipitation and temperature) and native vegetational cover in the evolution of soils, but there has been a tendency to minimise the importance of parent material, i.e. the nature of the underlying rock formations upon which the soil has formed. Consequently, soil investigations within the tropics continually sought to establish correlations between soil groups and the vegetational and climatic zones; less emphasis was placed upon identifying relationships between pedological and geological patterns.

This tendency has been countered by more recent studies which have aimed at re-establishing a truer perspective of the significance of lithological materials in determining the nature of soils. In Eastern Nigeria, the mineral properties of underlying rocks, together with the slope of the terrain, have been of paramount importance in resolving the broad characteristics of the main soil groups.

---

[1] This chapter on the pedology of Eastern Nigeria is based on a solicited manuscript from Dr P. D. Jungerius, a soil scientist formerly attached to the Ministry of Agriculture in Eastern Nigeria, now a lecturer in physical geography at the University of Amsterdam. The working paper has since been published separately as 'The Soils of Eastern Nigeria', *Publicaties van het Fysisch-Geografisch Laboratorium van de Universiteit van Amsterdam*, iv (1964), pp. 185–98.

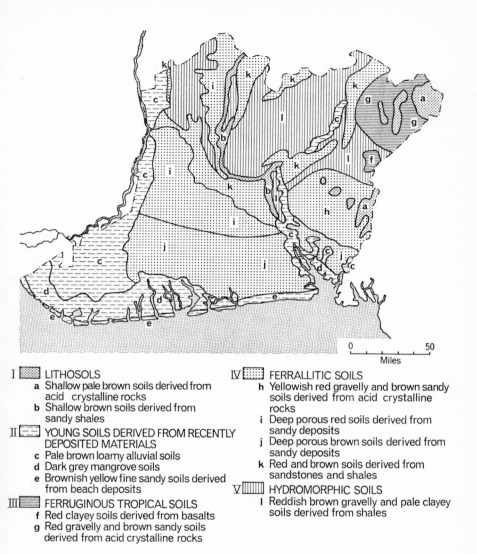

I ▨ **LITHOSOLS**
  a Shallow pale brown soils derived from acid crystalline rocks
  b Shallow brown soils derived from sandy shales

II ▱ **YOUNG SOILS DERIVED FROM RECENTLY DEPOSITED MATERIALS**
  c Pale brown loamy alluvial soils
  d Dark grey mangrove soils
  e Brownish yellow fine sandy soils derived from beach deposits

III ▤ **FERRUGINOUS TROPICAL SOILS**
  f Red clayey soils derived from basalts
  g Red gravelly and brown sandy soils derived from acid crystalline rocks

IV ▦ **FERRALLITIC SOILS**
  h Yellowish red gravelly and brown sandy soils derived from acid crystalline rocks
  i Deep porous red soils derived from sandy deposits
  j Deep porous brown soils derived from sandy deposits
  k Red and brown soils derived from sandstones and shales

V ▥ **HYDROMORPHIC SOILS**
  l Reddish brown gravelly and pale clayey soils derived from shales

FIG. 6.1   *Soils (after P. Jungerius)*

THE SOILS OF EASTERN NIGERIA

Fig. 6.1, of soils, which should be compared with Fig. 4.1, of geology, portrays the distributional pattern of five main *classes* of soils, differentiated on the basis of their morphology, also the degree of profile development

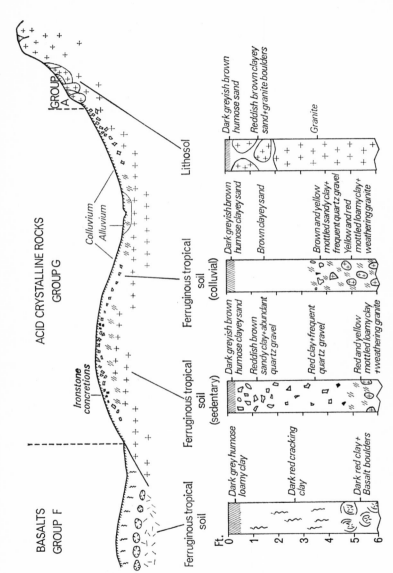

FIG. 6.2 *Soils formed over crystalline rocks (after P. Jungerius)*

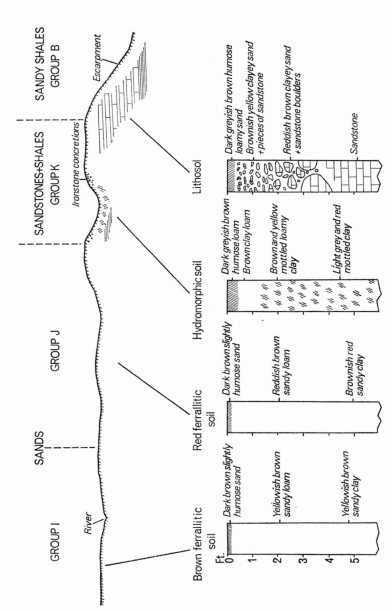

FIG. 6.3 Soils formed over various sedimentary rocks (after P. Jungerius)

(the soil profile is the appearance of the soil in a vertical cross-section from the uppermost, surface stratum or 'horizon' to the lowest one, that of the parent material).[1] Each class is divided in turn into *mapping units* according to characteristics of the sub-stratum or soil colour. The mapping units closely resemble soil *associations* or groups of soils geographically situated together in a pattern which may be repeated many times within an area, as in a catenary sequence. In Fig. 6.1, each map unit delineates an area which contains more than one soil group; only the dominant group in each unit is described in the following classification.

(i) LITHOSOLS

These are shallow, stony soils occurring on steep slopes where profile development is retarded due to erosion. They are formed over resistant rocks such as granites and gneisses, also sandy or silty shales.

(*a*) *Shallow pale-brown soils derived from acid crystalline rocks.* The soils of this group are found on the steep, rocky slopes of the hills and mountains in the area underlain by the granites and gneisses of the Basement Complex. They consist of pale-brown micaceous sand with abundant rock brash, overlying crystalline rock in various stages of decomposition (Fig. 6.2). Red and brown gravelly clay soils occur on the smooth slopes surrounding the steep uplands.

Areas of these soils are restricted to the more rugged sections of the Oban hills and Obudu Plateau and the steep topography is a limiting factor in their utilisation. Nevertheless they are often fertile soils due to the weathering of minerals in the parent material and a fair accumulation of organic matter from the rainforest cover, which breaks down more slowly than usual in the humid tropics due to the higher elevations and lower average temperatures.

(*b*) *Shallow brown soils derived from sandy shales.* Shallow, skeletal soils cover the steep escarpment slopes formed by resistant sandy shales and siltstones. These yellowish- to reddish-brown soils contain abundant, often ferruginised rock fragments, while unweathered rock is found at a depth of a few feet (Fig. 6.3). They are more fertile than those weathered to greater depths and this condition has led to cultivation of some areas despite the disadvantages imposed by steep topography. The principal area for these

[1] The classification followed in this chapter is adapted from J. D'Hoore, 'La carte des sols d'Afrique au sud du Sahara', *Pedologie*, x (1960).

soils is on the steep slopes of the Nsukka-Okigwi-Arochuku escarpment. In the Maku area, north-west of Awgu, these soils are intensively farmed by means of terraces – one of the most impressive systems of indigenous agriculture to be found within Eastern Nigeria (Part III, Chapter 10).

(ii) YOUNG SOILS DERIVED FROM RECENTLY-DEPOSITED MATERIALS

These are soils without well-developed horizons. They are derived from recent alluvium deposited by river or sea water and are subdivided into three mapping units.

(c) *Soils of the freshwater swamps; pale-brown loamy alluvial soils.* These soils are developed from sediments laid down in the northern section of the Niger Delta, also the extensive flood plain of the Niger and the more restricted one of the Cross River. Each year during the floods of the rainy season, fresh detrital material is being added, the coarse fraction to the natural levees and the fine fraction to the lower levels.

Although there is a range in texture from sand to clay, the soils are predominantly loamy. Profile development is restricted to the accumulation of organic matter in the surface horizon, and the formation of brown mottles in the generally pale-grey matrix (Fig. 6.4). Some swamp soils show a well-developed structure.

These fresh-water mangrove and *Raphia* swamp soils are potentially useful for the cultivation of wet rice and other crops but, except in the Anambra valley north of Onitsha, most of them remain in native swamp vegetation. Drainage control is the principal hurdle retarding their exploitation.

(d) *Soils of salt-water swamps; dark grey mangrove soils.* Much of the lower Niger Delta and the Cross River estuary consists of tidal mangrove (*Rhizophora*) swamps. The numerous creeks in these areas separate low-lying islands which are flooded daily with brackish or saline water. The soils are dark bluish-grey and acid silty clay loams. The more extensive older soils are covered with a thick layer of organic matter derived mainly from undecomposed mangrove roots (Fig. 6.4). The unfavourable natural environment has severely hindered the development of agriculture in this area.

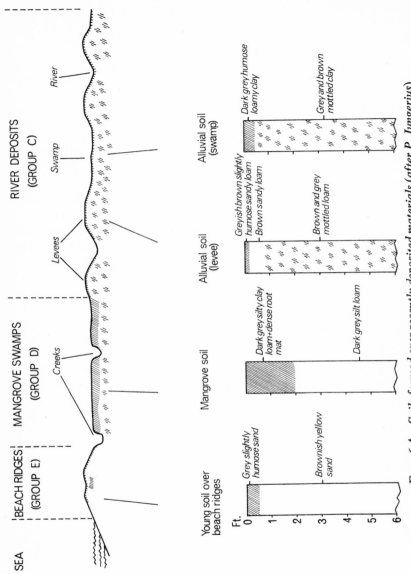

FIG. 6.4 Soils formed over recently deposited materials (after P. Jungerius)

(*e*) *Brownish-yellow fine sandy soils derived from beach ridges.* A strip of parallel sandy beaches and beach ridges, piled up by wave action, extends along the coast from Calabar to the western border. Profile development in this very young material is restricted. The excessively-drained soils on the ridges consist of deep, brownish-yellow coarse sand with a slightly humose grey topsoil (Fig. 6.4). Coconuts are sometimes grown on these sandy soils, and the Eastern Nigeria Development Corporation has established a small coconut plantation near Bonny. In the waterlogged depressions between the ridges, ground-water podzols occur, with about two feet of bleached sand overlying a dark brown organic pan.

(iii) FERRUGINOUS TROPICAL SOILS

The term 'ferruginous' refers to partially-weathered tropical soils which still contain a certain amount of weatherable minerals like silicates and sesquioxides (Fe- and Al-oxides). For this reason, they are sometimes also referred to as Fersiallitic soils (composed of *Fe*rrum, *Si*licum and *Al*uminium).

In Eastern Nigeria, these soils are rich in free iron and have a mineral reserve which may be appreciable, although aluminium is not abundant. They occur over basalts and acid crystalline rocks.

(*f*) *Red clayey soils derived from basalts.* These soils are developed over the scattered occurrences of volcanic rock in the vicinity of the upper Cross River, where the topography is gently undulating. Only the major area can be shown in Fig. 6.1. A dark reddish-brown, loam clayey surface soil grades into several feet of dark-red, structured light clay, with boulders of decomposing basalt at depth (Fig. 6.2). These soils have excellent physical properties and are well provided with nutrients by the weathering basic rock; economically, they are probably the most valuable soils in Eastern Nigeria and their limited extent is to be regretted. They are particularly suited to cocoa production, while maize and bananas also grow well on these basaltic soils.

(*g*) *Red gravelly and brown sandy soils derived from acid crystalline rocks.* These soils cover the northern outcrop of the Basement Complex in the planated zone surrounding the Obudu Plateau, comprising an undulating to rolling countryside with scattered steep-sided residual hills. The parent material is derived from coarse- to medium-grained granite and fine-grained gneiss, and the size of the sand grains varies accordingly.

The texture and colour of the soils change with the topographical site in a clearly-developed catenary sequence (Fig. 6.2). On the upper slopes, sedentary soils are developed with red to reddish-brown sandy and clayey surface layers containing abundant angular quartz gravel, passing down into a mottled clay layer showing traces of decomposing rock. At lower sites these soils become browner and more sandy, and there is often a narrow zone of concretionary ironstone. The soils on the lower slopes are developed in hill-wash material. They are brown and sandy, often with mottling in the sub-soil due to imperfect drainage conditions. Pale mottled sandy to sandy clayey alluvial soils occupy the narrow valley bottom.

These soils occur in sparsely-populated areas and remain largely under forest. Available plant nutrients are accumulated in the topsoil, the zone of weathering being often beyond the reach of plant roots. The agricultural value of these soils varies with the texture. Very concretionary upper slope soils and sandy lower slope soils have a low nutrient status and are liable to drought during the dry season. A high clay content improves the moisture and nutrient retention capacity.

(iv) FERRALLITIC SOILS

The term 'ferrallitic' (*Ferr*um and *Al*uminium) refers to completely weathered soils, i.e. soils in which the weatherable minerals have undergone complete desilification with a resulting weathered complex consisting only of the sesquioxides and resistant minerals such as quartz.

In Eastern Nigeria, these soils are rich in free iron but have a low mineral reserve and therefore a lower natural fertility than the ferruginous tropical soils. Covering over fifty per cent of the entire Eastern Provinces, they occur under a broad range of conditions.

(*h*) *Yellowish red gravelly and brown sandy soils derived from acid crystalline rocks.* The catenary sequence of the soils over the Basement Complex granites and gneisses of the Oban hills is similar to that described for group (*g*), except for differences caused by climate. Leaching and chemical weathering is more extensive due to the higher rainfall in this area and this results in somewhat paler colours and a lower nutrient status. Large stretches remain covered in high rainforest, cleared in only a few locations or large plantations of rubber and oil palm.

(*i*) *Deep porous red soils derived from sandy deposits.* These soils occur over the loose Coastal Sands which cover an extensive area in the west-central part of Eastern Nigeria; they are commonly referred to as the Red Earths or Acid Sands.[1] The topography is flat to undulating, with some ridges and deep and steep-sided valleys in the more northern edges. These soils are also present on the Nsukka-Udi Plateau, extending to the boundary with Northern Nigeria, where the terrain is rolling.

The effect of relief on these soils is limited and the morphology of the soil profiles is therefore remarkably uniform throughout the area. The soils consist of deep, porous, red to brownish-red, coarse sandy clays, with loose, reddish-brown coarse sandy upper layers. The topsoil is only slightly humose (Fig. 6.3).

The natural fertility of these soils is low as the minerals consist almost exclusively of quartz, iron oxides and kaolinite. Yet they are light and easily worked, and support some of the highest densities of rural population to be found anywhere in sub-Saharan Africa, under remarkable adaptations of bush-fallowing techniques and permanent compound cropping. (Part III, Chapter 10.) Principal crops produced include yam, cassava, beans and other vegetables, maize, bananas and oil palm fruits.

In an effort to explain the anomalous correlation of high population densities with poor soils (Part I, Chapter 2), it was suggested that the lengthy human occupancy of the areas with Acid Sands has led over the years to a marked deterioration in their productivity. While never of high fertility, widespread clearance of the former rainforest and disturbance of the delicate ecological balance between organic (humus) and inorganic matter in the upper soil horizons has undoubtedly produced the impoverished man-made earths witnessed in these areas today.

(*j*) *Deep porous brown soils derived from sandy deposits.* The soils in this group show much resemblance to the red ferrallitic soils of group (*i*). They are developed under similar conditions of parent material, but are paler in colour due to the higher rainfall in the area where they are found (southern Coastal Sands zone). Since the increase in the amount of rainfall towards the south is gradual, the boundary between the two groups is not sharp and the line in Fig. 6.1 should be taken as a cartographic device to portray a broad transition zone between the two mapping units.

The soils consist of a greyish-brown loose sandy to sandy loamy surface soil merging into a pervious brown sandy clayey subsoil at a depth of two to four feet. The humus content of the topsoil is low (Fig. 6.3). These soils are excessively drained but the dry spells are short. Heavy concentrations

[1] C. H. Obihara, 'The Acid Sands of Eastern Nigeria', *Nigerian Scientist*, i (1961), pp. 57–64.

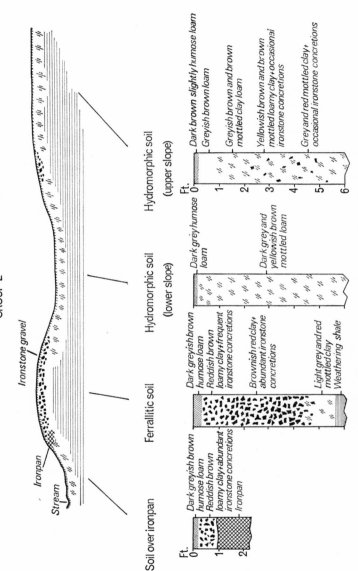

FIG. 6.5   *Soils formed over shales (after P. Jungerius)*

of population are supported in the south-eastern areas (Ibibioland) where the 'oil palm belt' attains its maximum expression.

(*k*) *Red and brown soils derived from sandstones and shales.* The soils of this complex group occur where shales are intercalated with sandstones in areas with undulating to rolling relief (Fig. 6.3). The uplands are usually overlain by coarse sandstones from which deep red sandy clay soils with a brownish-red, more sandy surface soil have developed. Brownish-red concretionary soils are occasionally found along the edges of the uplands.

The soils derived from shales on lower slope sites where the drainage is imperfect or poor show a dark reddish-brown, clayey surface soil merging into a strongly mottled light grey and red subsoil. Often, however, the valley bottoms are covered with sandy hill-wash material in which pale-brown and mottled sandy soils have developed.

The sandy soils are generally poor, but the nutrient-retaining capacity of the soils derived from shales is fairly high. Those derived from the argillaceous Bende-Ameki strata east of Umuahia, and the Eze-Aku Shales south of the Cross River between Ugep and Obubra, have proved particularly productive for cocoa and other cash tree crops.

(v) HYDROMORPHIC SOILS

The morphology of these mineral soils is influenced by seasonal water-logging caused by underlying impervious shales. The usually pale-coloured soils are mottled in the subsoil. Ferrallitic soils also appear in the areas covered by this mapping unit. The strong tendency to waterlogging has discouraged settlement and induced farmers to construct extra large hemispherical mounds or heaps in an effort to raise the roots of plants above the saturation line. In Abakaliki Province, mounds occasionally attain a height of over four feet, and a diameter of eight feet.

(*l*) *Reddish-brown gravelly and pale clayey soils derived from shales.* These soils are developed from the widespread, fine-grained sedimentary rocks in the northern section of the Cross River Basin. The area consists of an undulating to near-level plain with low ridges and wide shallow valleys. The network of stream valleys is dense due to the impervious nature of the bedrock, but only a few streams are perennial. These are often bordered by terraces of red sandy soils.

On the gently sloping uplands the soils consist of a brownish-grey humose loamy topsoil overlying a brownish-yellow to greyish-brown clay layer with brown mottling and weak blocky structure, which merges into a light grey and red mottled substratum at a depth of two to four feet (Fig. 6.5). Undecomposed shale is often encountered within a depth of six feet. Scattered ironstone concretions are a common constituent of these soils, particularly where the underlying shales contain silt or fine sand. While subject to waterlogging in the wet season, large quantities of yams, cassava, maize and some legumes are produced.

Brown and red gravelly soils, occasionally with ironpan, are found at the edges of the uplands and the summits of the narrow ridges, where erosion has removed the fine earth leaving the ironstone concretions as residue. These soils are better drained and have the characteristics of ferrallitic soils but are of lower fertility than the loamy upland soils.

In the depressions, where washed-out material is deposited into so-called rainfed inland swamps, loamy sand or loamy clay soils occur with grey and brown mottling below a dark-grey humose surface horizon. The natural fertility of the soils is low, because the shales are largely resistant to weathering; these swamp soils are widely used for rice during the rainy season, although water control measures are necessary to prevent variable water depths.

7 Soil Erosion and Deterioration

INTRODUCTION

WITHIN Eastern Nigeria there occur some of the most spectacular examples of soil erosion and 'badland' topography to be seen in West Africa. Erosion gullies attain a degree of severity and destructiveness seldom experienced in other parts of the continent. An unusual suscepti-bility to secular erosion, due to distinctive soils, geological formations and landforms (Figs. 6.1, 5.2, 4.1), together with marked disturbance of the natural vegetation cover by man in the course of agricultural pursuits, have initiated the regrettable scenes of devastation of natural resources in several areas of Eastern Nigeria today.

Dramatic gully erosion is most evident in the Plateau and Escarpment zone, particularly along the scarp of the Awka-Orlu Uplands and the Nsukka-Okigwi escarpment to the east (Fig. 7.1). Less pronounced though equally insidious sheet and gully erosion is widespread across the Eastern Provinces, however, extending from the plateaux in the north west as far south as the Coastal Plains, in the Ikot-Ekpene–Itu–Uyo triangle, and eastwards to the Cross River Basin. Soil deterioration and degradation, in terms of the progressive loss of the nutrients and breakdown of structure, is well-nigh universal, due largely to overfarming and primitive methods of cultivation.

Probably the most notorious area of gully erosion in Eastern Nigeria is near the villages of Agulu, Nanka and Oko in Awka Division, Onitsha Province.[1] Ten miles south of Awka, the headwaters of the obsequent North Awdaw River, a tributary of the northward-flowing Mamu (in the Imo Clay-Shale lowlands), have carved deep gullies and ravines into the

---

[1] The gullied areas in Awka Division, and others in Onitsha, Enugu and Owerri Provinces, were first studied in some detail by a geographer in 1948; the results of this investigation appeared in a monograph published by the Nigerian Government: A. T. Grove, *Land Use and Soil Conservation in Parts of Onitsha and Owerri Provinces.* Bulletin No. 21, Geological Survey of Nigeria (Zaria, 1951). (Hereafter referred to as *Land Use.*)

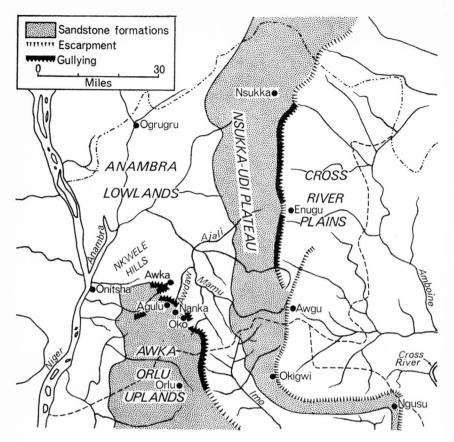

F IG. 7.1 *Gullied areas of the Nsukka-Udi Plateau and Awka-Orlu Uplands*

eastward-facing Nanka sandstone escarpment. The ferrallitic sandy soils or Red Earths proved highly vulnerable to erosion once their protective cover of moist evergreen forest was removed. Dissection of the Red Earth horizons exposes beds of unconsolidated pebbly, clayey and silty sands which were equally unresisting against the onslaught of mass wasting.

The resultant landscape today is catastrophic in its dimensions. Facing the observer on the scarp edge are the cliffed walls of a large erosion amphitheatre or complex of gullies. In the foreground, a gaping chasm fully a quarter of a mile wide and 350 feet deep has exposed highly-coloured layers of soil, sands and clays – deep red, yellow, white, pink, even violet in hue. Against these colours is set the dark-green foliage of relict patches of rainforest which have slumped into the ravines; elsewhere, the disintegrating soil has left vivid red smears down the gully walls.

Around the edges of the gullies, wide shrinkage cracks develop in the dry season, collecting run-off from the plateau surface in the rainy season. Sub-surface clay layers, lubricated by the ground water, induce earth slumping and slippage. Along the steep gully walls, outcropping interbedded layers of siltstones and ferruginised sandstones form resistant ledges, creating abrupt changes of slope. In the bottom of the gullies, small cliffs or 'knickpoints', often fifty feet high, indicate continued downward cutting as new base levels are sought.

Within the gullies, remnants of the original plateau surface in the form of spurs and outliers have side-slipped to lower levels. With the growth and convergence of adjacent gullies, the spurs shrink to narrow ridges, then to miniature *arêtes* or walls of sandstone with knife-like edges. Ultimately the gullies break through and the *arêtes* are reduced to a series of jagged, crumbling islands. The end result is a chaotic pattern of canyons, gullies and disintegrating sections of the former plateau.

Hardly less spectacular than the erosions in the Awka-Orlu Uplands are those found to the north and south of Enugu, along the pronounced scarp face of the Nsukka and Udi Plateaux (Fig. 7.1). From east of Nsukka to south-east of Udi, a succession of gullies has been etched into the False-Bedded Sandstones and Lower Coal Measure formations of the escarpment. The natural tendency towards geological erosion by headward-cutting obsequent streams has again been encouraged and accelerated by forest clearance on the plateaux for agriculture. A particularly extensive dendretic system of gullies is to be seen near Ukehe, where a mile-wide and 300-feet deep complex of ravines has produced a scalloped effect on the escarpment. The villagers of the area, having access to abundant land out on the lowlands to the east of the scarp, appear to be less concerned about the future of their plateau farmlands than those living in the Agulu-Nanka area. Indeed, according to Grove: 'the people are practically independent of the meagre yields the plateau lands can produce, and at Ukehe for example, where dissection is most thorough and complex, the greatest concern of the people is lest their cattle should fall into the ravines'.[1]

Around Enugu, the frequency of occurrence of erosion gullies led to the application 'Donga Ridge' in former times. On both sides of the steep descent to the town, via Milliken Hill, gullies are numerous and their heads

[1] Ibid., p. 39.

are threatening to cut the main Onitsha–Enugu road in places. Many thousands of acres of land are badly scarred in the immediate vicinity of the capital.

Five miles south of Enugu, the Nyaba river and its headwaters have created a further classic illustration of destructive erosion and land deterioration.[1] The gully zone or scarp course of the Nyaba comprises an intricate system of 'large, digitate gullies with vertical-walled "bulbous" heads'.[2] V-shaped in profile, the gullies fall steeply from the plateau surface and converge to form deep canyons which in turn unite to create a broad major trench at base level. At present, most of the uppermost gullies are inactive, and headward cutting with rapid enlargement of ravines has apparently ceased; side-slipping and slumping of material still takes place, however, particularly during severe storms in the rainy season. The slowdown in erosion may be partially attributed to conservation measures instituted by the Forestry Department in 1922, when the Udi Forest Reserve was established around the headwaters of the Nyaba river, and farming in the immediate vicinity of the gullies was prohibited. On the other hand, the presence of 'knickpoints' or marked breaks of slope in the gully bottoms suggests that conditions are by no means stable and renewed erosion may easily be initiated.

The lower valley or plains course of the Nyaba reveals an aggraded and braided stream bed, with a confused pattern of sedimentation. Accumulation of yellow and orange sterile sands has been so great that tributary streams have been unable to join the main river and small lakes have resulted from their damming. The only benefit to be derived from the lower Nyaba and other aggraded rivers near Enugu is an unlimited supply, conveniently to hand, of large quantities of building sand for the booming construction programme within the urban area.

THE EVOLUTION OF THE GULLIES

Attention may now be directed towards reconstructing the sequence of events which has led to the creation of erosion gullies in the areas described above, with a view to identifying more precisely the principal factors and agents responsible for these disturbing examples of 'badlands' in Eastern Nigeria. Measures taken thus far to check the progressive advance of the

[1] This area has been authoritatively investigated by Jean Carter, 'Erosion and Sedimentation from Aerial Photographs: a Micro-Study from Nigeria', *The Journal of Tropical Geography*, xi (1958), pp. 100–6.

[2] Grove, *Land Use*, p. 42.

gullies and to slow down the rate of soil deterioration will also be reviewed.

A study of the evolution of the erosional features of the upper Awdaw valley around Agulu-Nanka will serve as a useful case-history for interpreting comparable landscapes elsewhere in the Awka-Orlu Uplands, and on the Nsukka and Udi Plateaux and Escarpments further east.

The gullied hill slopes south of Awka were first commented upon by European officials in the early years of this century, so that the devastation is not of recent origin. Nevertheless, growth of some of the gullies is said to have been very rapid following continued destruction of woodland within the last fifty to sixty years. The upland areas of the Nanka Plateau have long been favoured for settlement since their light sandy loams are easy to cultivate with simple, short-handled hoes and, in contrast to the heavy clay soils of the adjacent Mamu lowlands, they are free-draining. Ease of movement for social intercourse in earlier times was also assured. The water-table was high and the life-giving liquid was obtainable from springs at many places. Thus communities flourished and population pressure on the land increased, requiring in turn more extensive deforestation and utilisation of the soil for farming. Densities in excess of one hundred persons per square mile required supporting under traditional methods of shifting cultivation and 'bush' fallowing. By perhaps 1850 the pattern was set for triggering off the sequence of events which has created the ravaged scenes of the present day.

With the clearance of more land for agriculture within the scarp zone, and the encroachment upon areas of increased slope (with gradients varying from 5° to 30°), the sandy soils were exposed more directly and continually to the climatic elements, and accelerated soil erosion was precipitated. Unprotected by the deep shade of trees, surface temperatures now fluctuated markedly in response to diurnal variations of air temperature, particularly pronounced in the dry season. The high relative humidities of the climate within the former rainforest (due to lower temperatures and continuous transpiration) were replaced by the lower humidities of a grassland climate. The combined effect of high day temperatures and greater aridity was to increase the rate of chemical changes in the soil and to retard the decomposition of organic matter. In sum, the changing atmospheric conditions were detrimental to plant regeneration and accumulation of organic material.[1] Given the attitudes, objectives and technical abilities of the people, degradation and destruction of the soil were unavoidable concomitants of the mounting population pressure upon deteriorating physical resources.

[1] Ibid., p. 19.

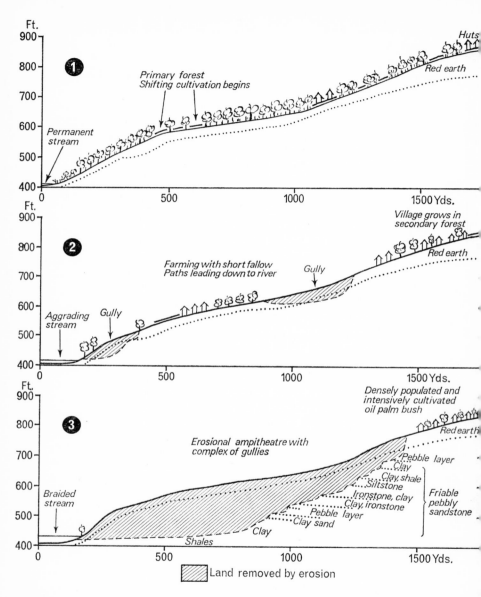

FIG. 7.2   *Cross-section of Agulu-Nanka gullies depicting their evolution*
(*after A. T. Grove*)

The Agulu-Nanka area, in common with much of central Eastern Nigeria, has an average annual rainfall of around 80 inches a year, and a rainy season of seven to eight months' duration. Regular and often intense precipitation from May to October, bearing down on the exposed soil surfaces, soon added its destructive force to that of the other climatic elements. Excessive leaching and wind-erosion of the soil horizons were followed by wholesale removal of the topsoil through sheet wash and gullying, which initiated and promoted the growth of ravines on the scarp face. Eventually the underlying rock formations of unconsolidated and friable sands were exposed and mass earth movements through slumping, sliding and downhill 'creep' were set into motion, adding to the growing derangement of landscape features (Fig. 7.2).

It is worth observing that the pace of erosion proceeded irregularly, even spasmodically, and continues this way at the present time. Violent storms with short yet torrential downpours, common towards the beginning and end of the rainy season, can inflict much more damage in the space of an hour or so than regular rains over a much longer period of time. Of even greater importance are the rare storms of catastrophic proportions which have a destructive force far in excess of that mounted over several years of more normal weather experience.

Apart from periodic clearing of the forest, through burning and slashing by matchets, other cultural practices of the farmers did nothing to alleviate the increasingly hazardous situation. These time-honoured practices are frequently still witnessed today. The perennial use of fire for clearing the undergrowth, flushing out small game, and encouraging a green 'bite' for livestock towards the end of the dry season, causes further deterioration of the soil structure and loss of organic matter. The thin layer of ash from the fires, while potentially useful to crop growth, is swiftly removed by the rains. The traditional construction of hemispherically-shaped mounds or heaps for yams and cassava tends to concentrate run-off into rills which increase in size down-slope; the mounds are seldom built on the contour. Mud walls around compounds and gardens are similarly ill-aligned for conservation purposes. There are extensive areas of bare ground around homes, along footpaths and bicycle trails (frequently up and down hill) and at market places. These serve as catchment areas for storm water run-off. Grove estimated that, in 1948, some 5 per cent of Oko village land consisted of bare compound land, market places, roads and footpaths.[1]

---

[1] Ibid., p. 27.

Once the erosion gullies had appeared, communities were powerless to check their growth, if indeed any desire arose, or efforts were made in the early stages, to arrest their development. Lowering of the water-table, landslides around deforested spring-heads and the aggrading of stream beds were, in addition to loss of farmland, the greatest inconveniences suffered by the population, necessitating longer journeys in search of water and frequently a climb down into the nearest gully to gather water from its base. The daily movement of innumerable human feet, across the adjacent slopes and up and down the gully sides, created additional channels for the concentration and accumulation of surface water, the erosive agent most responsible for the growth of the gullies.

THE HISTORY OF SOIL CONSERVATION MEASURES

Only with the formation of government departments responsible for promoting forestry and improvements in agriculture was serious attention directed towards trying to control the gullies and to maintain or restore soil fertility in those areas where degradation was clearly a matter of rising concern. A Department of Agriculture was established in Eastern Nigeria in 1910 and among some of the earliest experiments undertaken were soil conservation trials aimed at maintaining productivity under various crop treatments and land-use procedures. Later, with the formation of a Forestry Department, the first efforts to check gullying were undertaken, involving the reafforestation of exposed slopes and conservation of relict woodland in critical areas. These early programmes were highly localised, however, being restricted to a few areas in the old Onitsha Province, and no overall conservation scheme was evolved for the region as a whole.

Immediately prior to the Second World War, in 1938, a directive from the British Colonial Office drew the attention of the Nigerian Government to the dangers of uncontrolled soil erosion; but the onset of the war precluded any determined effort to face up to the problem. In 1943 the Secretary to the Eastern Provinces empowered Native Authorities in affected areas to pass legislation relating to soil conservation, but such Orders which resulted were too limited in scope and difficult to enforce.

Only at the end of the war was a more concerted effort made by the Government to tackle land deterioration and gully formation, using Colonial Welfare and Development funds. In 1945, a soil conservation scheme was initiated at Agulu which aimed at checking the growth of the gullies and to serve as a demonstration area for farmers from other eroded areas. The experiments undertaken and measures adopted were based largely upon American experience in combating erosion in the southern U.S.A., also anti-erosional work pursued elsewhere in British Africa (e.g., Kenya and Southern Rhodesia).

An effort was first made to prevent surface water from directly entering the heads of the gullies. Water courses, drainage ditches, sunken footpaths and other catchment features were blocked and the water diverted. Circular soak-away pits or sumps (four feet in diameter, five feet deep) were dug alongside footpaths to encourage run-off to disappear underground. A less orthodox technique known as wave-bedding was tried out around the heads and sides of the gullies. This involved 'pock-marking' the surface of the land with shallow, saucer-shaped holes (five to six feet in diameter, surrounded by a one-foot earth ridge) in a continuous pattern. Rainwater collected in the depressions and was thus prevented from running down the slopes. The remedial effect of wave-bedding on the soil is similar to that achieved by basin-listing or cross-tying of contour ridges in the U.S.A.

Shrubs and trees were planted in the holes to stabilise the land. Establishing a protective plant cover within the watershed was considered absolutely vital. Farming was prohibited in the areas immediately adjacent to the gullies and a number of shrubs and trees were introduced, capable of surviving on the dry, acid sands and sterile bare slopes. Among the plants tested were the roseaceous shrub *Acioa barteri*, which acts not only to protect the soil from raindrop erosion, but also probably to restore fertility (as a by-product, its branches make useful sticks for supporting the yam vine); the cashew tree *Anacardum occidentale* which grows readily on sandy soils, yields a deep leaf litter, and provides a nut of commercial value; oil bean trees; bamboo and various grasses. These plants were set out not only on steep slopes around the gullies but within the ravines themselves.

Mechanical works in the form of earth-check dams across the gully floor, and stepping of walls, were also undertaken, with less favourable results. Some dams were swept away by flood waters from violent storms; others were undermined or ruptured by the steady pressure of backed-up water. Beyond the ravines, on gently-sloping land where farming might be permitted, narrowed-based contour terraces (bunds) were made obligatory, and strip cropping was strongly encouraged.

All these methods of controlling erosion were shown, in aggregate, to be capable of slowing down the pace of destruction and stabilising the smaller gullies. Soil regeneration in peripheral areas under introduced vegetation was a more questionable matter, owing to the relatively short period of the trials. Unfortunately, the whole Agulu scheme was curtailed in 1948. There was little indication of willingness on the part of the mass of the farmers to adopt the procedures deemed necessary to heal the land; in fact, a marked measure of apathy was noticeable. Attempts to enforce farming on the contour, using ridges for growing yam and cassava in place of the familiar heaps, were viewed with suspicion, if not hostility. New methods of utilising the soil, involving systematic crop rotations, the planting of cover crops, mulching, green manuring and the addition of chemical fertilisers were similarly rejected by the rural population at large, being strange and untested innovations and beyond the technical and financial means of the ordinary farmer.

The problems of promoting conservation in Eastern Nigeria are, of course, exacerbated by the excessive pressure of population on the land, which has led in turn to a marked sensitivity to matters involving land tenure. Violent and prolonged land disputes between villages are still a commonplace experience in many areas of the country; these controversies over ownership of land and allocation of farming rights paralyse any effective scheme for soil conservation over a watershed area.

Other sociological, even religious, considerations have hampered the work of soil conservationists. The destructive gullies have at times been viewed as the handiwork of offended gods whose appeasement involves the offering of food, libations and other ritualistic ceremonies, not alien-directed attempts to confine and control the erosion, which can only bring down fresh vengeance upon the recalcitrant community.

Returning ex-servicemen from the Second World War, dissatisfied with governmental handling of their rehabilitation into civilian life, took the opportunity of embarrassing the authorities by urging that it was the responsibility of the government to provide sizeable sums of money for the purchase of cement with which the gullies could be arrested. Elsewhere, the people suspected that geologists had located mineral wealth within the gullies and wanted to acquire the land by force from the local community.

The phenomenon of peasant conservativeness and hostility *vis-à-vis* recommended changes in land-use techniques and tenurial patterns is by no means unique to Eastern Nigeria; it has been the widespread experience of agriculturists the continent, indeed the world, over. Nevertheless, the dangerous situation in the Eastern Provinces was hardly matched in other areas of Africa and the need to win the co-operation and good will of the farming community has rarely been so great elsewhere on the continent.

Since 1950, very little work aimed at controlling the Agulu-Nanka gullies has been attempted. Local authorities in a rather desultory fashion continue to sponsor the construction of earthen dams, sink pits and contour banks from time to time. There is little indication, however, of any widespread enthusiasm for stopping the erosions, or adopting improved farming methods which might alleviate the risk of further gully encroachment. Limited experiments in cover planting both in and around the gullies were initiated in 1962 by an American soils adviser, using a tropical leguminous vine, *kudzu*, Brazilian lucerne (*Stylosanthes gracilis*), and indigenous grasses such as molasses grass and the so-called giant star Bermuda grass. Preliminary results revealed that fertilisers are necessary for establishing these plants firmly and seeing them through the dry season. The costs of rehabilitating the erosions may be increased inordinately in view of this requirement.

Periodic statements from the Ministry of Agriculture declare the Government's intention to take up once again the acute problem of soil erosion in Eastern Nigeria, particularly in Awka Division. The declarations meet with the approval of educated community leaders and local government officials; but, in view of the recent history of unsuccessful attempts to persuade farmers to adopt voluntarily the necessary steps for combating erosion, it is doubtful whether the mass of the people will respond to the call for necessary action unless firmer and sounder procedures are adopted.

While vigorous efforts to control the gullies of the Awka-Orlu Uplands and Nsukka-Udi Escarpment ought to be reinstituted, the spectacular chasms should not be allowed to monopolise the attention of the soil conservation services. Soil deterioration through sheet-washing and cultural malpractices on the part of land-hungry farmers is continuing over considerable areas of the Eastern Provinces. Further dissection of the land and more gullying may be imminent, jeopardising the future of agriculture in many places. The way in which the authorities come to grips with the total agro-socio-economic problem of soil erosion in all its forms will determine the destinies of future generations in Eastern Nigeria.

8 Meteorological and Climatological Considerations

INTRODUCTION

OUR study of the Eastern Nigerian environment is enhanced by an appreciation of the spatial arrangement and temporal variability of the elements of weather and climate. Atmospheric conditions from day to day, from month to month and from year to year are all-pervasive in their impact upon both earth and man. Weather and climate have a direct effect on human health, efficiency and psychology, and an indirect effect on man's economic activities through the interaction with soils, vegetation, crops, geomorphic processes and hydrology.

The geographer attempts to establish and measure environmental parameters, the atmospheric metes and bounds which support or retard man's efforts to maintain life and advance his society. The settlement which is negotiated and finally established between an environment and a technology lies at the very core of geographical studies.[1] As in so many other branches of geographical enquiry, there have been outstanding changes in the concepts of meteorology and climatology in recent years. The trend has been away from measurements and interpretation of simple parameters such as temperature, rainfall and relative humidity, towards the measurement of energy and moisture exchanges or 'fluxes'. Instead of mere description of atmospheric conditions – synoptic meteorology and climatology – the emphasis is on achieving an understanding of the heat and water balance at the earth's surface.

It is now clear that environmental climatology of all sorts – ecological, physiological, hydrological and so on – must increasingly depend on the study of exchange processes. It is not air temperature that is *per se* significant; it is the heat exchange that occurs at the leaf, sea, soil or skin surface. Rainfall alone is not enough; we have to consider

---

[1] For a development of this theme, see P. W. Porter, 'Environmental Potentials and Economic Opportunities – A Background for Cultural Adaptation', in W. Goldschmidt *et al.*, 'Variation and Adaptability of Culture: A Symposium', *American Anthropologist*, lxvii (1965), pp. 409–20.

the evaporation losses also, again off leaf, sea, soil or skin. Carbon dioxide and ozone exchanges similarly protrude themselves, as does the fall of contaminants from the atmosphere to the surface. Moreover, the effective climate is not merely atmospheric; it extends deep into the soils and the oceans.[1]

Much as one would like to present the climate of Eastern Nigeria within the framework of the 'new school' of climatologists, it is not possible at present to do so since virtually none of the field measurements necessary for achieving such an analysis has been undertaken. A more orthodox, descriptive approach has therefore been followed.

The climatic characteristics of Eastern Nigeria are best understood in terms of a number of different weather 'complexes' experienced in the course of a normal year. These consist of a dry season, a north-easterly air stream (the Harmattan), a south-westerly 'monsoon' air stream, squall lines or disturbance lines, thunderstorms, steady rain and drizzle, and a 'little dry season'. Their time of occurrence and their duration, also the intensity with which the phenomena are developed – these are the basic components of Eastern Nigeria's climate.

To obtain a fuller picture of the sequence of weather and climate over a year, a month-by-month descriptive and interpretative account of atmospheric experiences is given.[2] It is supported by descriptions of actual weather experiences in Eastern Nigeria for the year 1965, with the reservation that conditions in any one year may well be atypical, particularly in their day-by-day details.[3]

JANUARY

General Conditions. During the northern hemisphere winter, an extensive anticyclone (the Saharan anticyclone) establishes itself at the surface, approximately along latitude 20° N (the northern Niger Republic). On its southern side an easterly air stream, the Harmattan (known at sea as the 'trade winds'), extends its influence progressively towards the Equator, reaching the Gulf of Biafra in early January. The Harmattan is progressively weakened in its southward penetration and rarely affects the coastal

[1] F. K. Hare, 'The Concept of Climate', *Geography*, li (1966), p. 99.
[2] This account is based on a description of the 'weather year' over West Africa contained in Section I of S. Gregory, *Rainfall over Sierra Leone*, Department of Geography, University of Liverpool, Research Paper No. 2 (Liverpool, 1965), pp. 1–16. (Hereafter referred to as *Rainfall.*)
[3] Weather descriptions for Eastern Nigeria are extracted from the *Agro-Meteorological Bulletin* of the Nigerian Meteorological Service, Lagos.

stretches of the Eastern Provinces for more than a few days each year.

Emanating from the dry continental tropical air mass (cT) over the Sahara, the Harmattan is similarly a dry and dessicating wind, laden with dust, which severely reduces visibility in the areas over which it blows. The onset of the Harmattan is characterised by a sudden drop in atmospheric humidity, which may fall from 85 per cent relative humidity (R.H.) to 30 per cent in a few hours (Table 8.14). A marked backing of the wind from 270° to 090° several hours before the humidity falls, together with increased wind velocity during the first day or two of the Harmattan season, are also typical features. The reduction in visibility or 'Harmattan haze' is a result of dust particles trapped in the lower levels of the atmosphere by a strong temperature inversion within the tropical easterlies. 'This marked stability of the atmosphere also ensures that rainfall is effectively absent, so that a period of intensive potential evapotranspiration (because of low humidity) coincides with an absence of incoming moisture.'[1]

From around the middle of January onwards, the area within Eastern Nigeria under Harmattan influences slowly but steadily contracts, as a shallow tongue of moist oceanic air moves in from the south-west to replace it at ground level. This air stream – sometimes referred to as the anti-trades – emanates from the maritime tropical air mass (mT) over the Southern Atlantic. As a result of the great land mass north of the equator between 15° W and 10° E, and a corresponding absence of land south of the equator between the same meridians of longitude, a monsoonal inflow of air occurs in West Africa and Eastern Nigeria long before the vernal equinox (21 March).

The zone of contact between the cT air mass and the Harmattan stream to the north, and the mT air mass and the monsoon air stream to the south, is called the Inter-Tropical Convergence Zone (I.T.C.Z.). Formerly referred to as the 'Inter-Tropical Front', this name is now discarded since it implies a rather precise line of contact and associated frontal disturbances similar to those experienced along the Polar Front in intermediate latitudes. In fact the passage of the I.T.C.Z. is often unaccompanied by any marked change of weather in terms of cloud sequence or precipitation. In Nigeria, the I.T.C.Z. is usually called the Inter-Tropical Discontinuity (I.T.D.); this title emphasises the contrast in the characteristics of air masses on either side of the zone, rather than their tendency to approach each other from dissimilar areas of origin and to grow more alike in the zone of convergence.

[1] Gregory, *Rainfall*, p. 2.

JANUARY 1965

Actual Conditions. With a broad-scale Upper Westerly trough dominating North Africa (the Sahara) during the early days of the month, the Inter-Tropical Discontinuity moved from about 9° N (over the 'Middle Belt' of Nigeria) to 12° N (around Kano) during the first five days; local thunderstorms (L.T.) affected the coastal areas and other parts of the region during this period, due to convectional heating of moist mT air over the land, and unseasonal rainfall was received at Enugu. On 11 January an easterly wave moved into Eastern Nigeria; it produced no significant weather but appears to have been responsible for a southward movement of the I.T.D. to 8° N (over the Benue Valley) that day. The I.T.D. returned gradually to 12° N by the fifteenth but a second easterly wave on the seventeenth brought it southwards again to 8° N. There was little movement of the I.T.D. during the rest of the month. The period 19 to 22 January saw the re-establishment of strong anticyclonic conditions over the Sahara which dominated the weather to the south for the rest of the month.

Extensive Harmattan haze moved into the northern areas of the Eastern Provinces on the twenty-first and fluctuated in intensity for the next several days. In the coastal areas the thunderstorms of the first week gave way to more settled conditions during the second week, although early-morning fog was frequent. Generally unstable conditions returned in the third week, giving rise to local showers and thunderstorms in the extreme south. Scattered thunderstorms continued to affect the coastal areas, particularly the Niger Delta, until the end of the month.

TABLE 8.1

Climatic data, January 1965

Station	Mean Temp. (°F)			R.H. (%)	Rainfall			
	Daily Mean.	Max.	Min.		Inches	Dept.fr. norm	Rainy days	Dept. fr.norm
Enugu	80	91	71	77	4·6	+4·0	2	+1
Port Harcourt	79	88	71	88	1·5	+0·4	5	+2

FEBRUARY

General Conditions. By mid-February the shallow monsoonal inflow from the south west has asserted itself and most of Eastern Nigeria is under its influence. The I.T.D. is fairly consistently north of the region, between 10° N and 12° N. Along the zone of contact between the dry easterlies to the north and humid westerlies to the south, it is the dry air that is strongly subsident, so that instability within the humid westerlies is inhibited not only at the I.T.D. but also for as much as 200 miles south of it (i.e. over the greater part of the Eastern Provinces). Thus the period of rainless, atmospherically dry conditions of the Harmattan is followed by an almost equally rainless, although atmospherically humid, period. Occasional showers (even heavy rain) may occur during these conditions, especially over the Eastern Highlands, but they are not of frequent occurrence.

FEBRUARY 1965

Actual Conditions. During the first week of February, the weather was dominated by a high-pressure area over the Sahara, easterly winds prevailed over much of Nigeria and the I.T.D. was pushed southwards to about 7°N, even reaching 6°N (south of Onitsha) on the sixth. Then, under the influence of a pronounced trough in the Upper Westerlies over Nigeria, the I.T.D. moved to 14° N on the eleventh. Unsettled weather conditions spread to nearly all parts of the Eastern Provinces on the tenth and eleventh, with monsoonal-type rain and thunderstorms.

The I.T.D. moved from 14° N to 12° N on the fourteenth (where it lay for the rest of the month) as the trough cleared Nigeria and proceeded eastwards, while pressure increased again over the Sahara. A low-pressure area over the Cameroons hindered anticyclonic easterly winds from penetrating Eastern Nigeria, and fairly frequent thundery outbreaks continued within the humid maritime air mass, particularly in the Delta area. After the twentieth thundery activity decreased and from the twenty-fifth onwards the weather was settled over the entire region.

TABLE 8.2

Climatic data, February 1965

Station	Mean Temp. (°F)			R.H. (%)	Rainfall			
	Daily Mean	Max.	Min.		Inches	Dept.fr. norm	Rainy days	Dept.fr. norm
Enugu	81	92	71	71	0·9	− 0·3	1	− 1
Port Harcourt	81	90	71	84	2·8	+ 0·6	8	+ 1

MARCH–APRIL

General Conditions. During these months, the largely rainless but atmospherically humid phase is increasingly interrupted by showers and thunderstorms, with considerable and spectacular lightning. These storms may appear to be distributed at random but they are invariably associated with squall lines or disturbance lines (D.L.) which travel in a general east to west direction against the prevailing surface air stream and extend north to south for distances of between fifty and two hundred miles.

> Their development is related to a combination of surface land features, the characteristics of the low-level westerlies, and the existence of the tropical easterlies above. The depth of the monsoonal westerlies tends to increase slowly southwards from the ITCZ. Once they are between 2000 ft and 4000 ft. thick, their high humidity and their basically unstable nature lead to the development of cumuliform clouds. The inversion of the easterlies above, however, normally prevents air rising upwards beyond the top of the westerlies. In areas of marked, though not necessarily high, relief forms sufficient additional instability can be imparted for limited showers to develop.[1]

More violent weather, in terms of disturbance line thunderstorms (commonly but erroneously termed 'tornadoes' by laymen in Nigeria) and heavy precipitation, takes place when the stabilising effect of the overlying easterlies is temporarily but firmly replaced by conditions of instability.

> The most likely cause of this is the occurrence of moving wave forms within the easterlies, these leading to rough north–south belts of extreme convergence in their lower layers. The vertical coincidence of such an unstable layer above a surface of intensified instability in the westerlies can lead to rapid thunderstorm development; furthermore, as any such waves would tend to move westwards with the easterly airstream, the westward movement of these disturbance lines against the surface airstream becomes explicable. Also, the influence of surface relief tends to lead to areas of preferential development, while movement of the storms away from such areas leads to their ultimate diminution and dying away.[2]

In Eastern Nigeria, the Obudu Plateau and the Oban hills (also the Cameroon mountains east of the region) play a dominant role in providing the 'triggering effect' necessary to initiate wave forms in the upper air streams; the resultant squall lines then travel swiftly (twenty to thirty miles per hour) over the Cross River Basin westwards to the densely populated zone of Plateaux and Escarpments and the Coastal Lowlands, at times inflicting considerable damage on property and crops, before dim-

[1] Gregory, *Rainfall*, p. 10.   [2] Ibid.

inishing and ultimately dying away. With the passage of a squall line at any particular place, steady rain or drizzle succeeds the thunderstorms for several hours. It is also likely that the Nsukka-Udi Plateau favours the initiation of such storms, when other conditions are favourable. As surface heating is an additional factor in inducing instability, late afternoon to early evening is a common time for the occurrence of these thunderstorms.

It is worth emphasising that these weather disturbances do *not* occur along or within the I.T.C.Z. or I.T.D. but well south of it (from 200 to 400 miles), beyond the belt of shallow, rainless but humid air, to the point where the monsoonal westerlies have become deep enough to permit the initial development of instability. At times a closer relationship between storms and the I.T.D. is recognisable but the disturbances are usually less violent.

MARCH 1965

Actual Conditions. During the first week, the general situation over the Sahara and North Africa was weakly anticyclonic, and the I.T.D. was fairly steady at 10° N. The weather in Eastern Nigeria was generally settled except for scattered evening thunderstorms in the south. These conditions gave way to more unstable conditions, with the westerly air stream advancing mT air northwards to a zone of contact with cT air at around 12°N. Thundery activity was frequent and often widespread in the Eastern Provinces from the seventh to the nineteenth and minor line squalls were experienced.

For the remainder of the month there was a gradual re-establishment of anticyclonic conditions over the northern part of Africa and the I.T.D. retreated as far north as 7° N (over Nsukka) before establishing a steady position between 9° N and 10° N until the end of the month. A trough in the Upper Westerlies in the period 24 to 28 March had little effect on the weather but, following its passage eastwards, an area of instability moved over Eastern Nigeria to the mid-west on the twenty-eighth and there was a marked increase in the strength of the easterlies.

TABLE 8.3

Climatic data, March 1965

Station	*Mean Temp.* (°F)			*R.H.*	*Rainfall*			
	Daily Mean	*Max.*	*Min.*	(%)	*Inches*	*Dept.fr. norm*	*Rainy days*	*Dept.fr. norm*
Enugu	84	94	73	70	3·0	−0.9	5	−1
Port Harcourt	83	91	72	80	5·8	0	10	−3

APRIL 1965

Actual Conditions. A trough in the Upper Westerlies affected Eastern Nigeria between 5 and 8 April and caused the I.T.D. to move northwards to 14°N. This brought unstable conditions to the region but by the ninth the trough had moved away eastwards and a general improvement in the weather took place as the I.T.D. retreated southwards to 10° N.

From the eleventh to the twentieth, there was a progressive seasonal movement northwards of the anticyclonic belt in the upper levels and, as the easterly current deepened, waves in that current moved westwards over the region. The I.T.D. advanced to 17° N and thundery activity was frequent over almost all parts of the Eastern Provinces.

A fairly deep current of moist south-westerly air continued to build up during the rest of the month, and squall-line thunderstorms affected most parts of the region.

TABLE 8.4

Climatic data, April 1965

Station	Mean Temp. (°F)			R.H. (%)	Rainfall			
	Daily Mean	Max.	Min.		Inches	Dept.fr. norm	Rainy days	Dept.fr. norm
Enugu	82	92	74	73	8·8	+1·9	10	0
Port Harcourt	81	89	72	83	9·1	+1·9	12	−2

MAY–JUNE–JULY

General Conditions. The month of May is usually considered to mark the formal opening of the rainy or wet season. The I.T.D. is now firmly established well to the north of the Eastern Provinces and may reach latitudes of 15° N to 20° N over the Sahara in succeeding months. The belt of squall lines or disturbance lines also migrates northwards and, to the south of it, the thunder-showers are replaced by steady rains or drizzle. Unlike the short, sharp and often violent downpours associated with the D.L.s, the wet season rains are longer and steadier, sometimes prolonged for up to seventy-two hours continuously although they are more normally of some six to eighteen hours duration. The intensity of precipitation may not be as great as from the thunderstorms but the cumulative effect is much greater.

Temperatures are at their lowest for the year while the R.H. is high. The sky is overcast for much of the time and sunshine is at a premium. Nevertheless, it is incorrect to suggest that conditions are uniformly dull or

E   F.E.N.

monotonous. Bright spells occur when the rains cease, the clouds clear, and sunny humid conditions are experienced. There may in fact be considerable variety in the day-to-day and even hour-by-hour weather experiences in Eastern Nigeria, both during the rains and in the dry season, which confounds the climatological generalisations of the humid tropics based on means and norms.

At the height of the rainy season in June and July, and again in September, region-wide variations in the pattern of steady rains are related to rain belts, some 200 miles across, which move inland from the Gulf of Guinea and are interspersed with areas of clearer skies. 'The immediate differentiating factor would seem to be the depth of the westerlies. When these exceed 5,000 to 6,000 ft. in depth, rainy conditions are likely; when they are slightly shallower then dry days intervene'.[1] The lessening in depth of the westerlies and their greater stability is believed to be related to Atlantic anticyclonic conditions in the southern hemisphere.

MAY 1965

Actual Conditions. For the first half of the month there was a gradual deepening and strengthening of the Easterly Trades and the I.T.D. fluctuated between 13° N and 15° N as a succession of mainly weak waves in the Easterlies moved westwards across the region.

A line squall affected the southern coastal and Delta areas on the eleventh. Seasonal thunderstorm activity continued during the second half of the month. Major line squalls which developed east of Yola in the Cameroon massif moved westwards across Northern Nigeria on the twenty-fourth, twenty-sixth and twenty-seventh, producing only side-effects in the Eastern Provinces, but a line squall on the twenty-seventh swept directly across the region. Again on the twenty-ninth a line squall moved from the Benue Valley south-westwards to the Delta area.

TABLE 8.5

Climatic data, May 1965

Station	Mean Temp. (°F)			R.H.	Rainfall			
	Daily Mean	Max.	Min.	(%)	Inches	Dept.fr. norm	Rainy days	Dept.fr. norm
Enugu	81	90	73	76	5·4	− 2·4	11	− 4
Port Harcourt	81	88	72	82	7·7	− 1·8	15	− 2

[1] Gregory, *Rainfall*, p. 11.

JUNE 1965

Actual Conditions. The marked northward movement of the axis of the sub-tropical anticyclone at all levels was a good indication that the region was well into the rainy season. This northward movement gave way to the moist monsoon air stream, the major feature of this season. On the surface the position of the I.T.D. which was at 13° N early in the month moved north to about 21° N and then finally settled at 19° N by the end of the month. The weather experienced during the month was influenced by the intensification of the sub-tropical anticyclones alternatively in the northern or in the southern hemisphere, in that this established the depth of the monsoon air over the country.

In Eastern Nigeria, rainfall was generally above average (by 10 per cent to 60 per cent) except for isolated areas around Aba, Afikpo and Calabar where it was below average by 10 per cent to 40 per cent. Rainfall values of 8 inches to 10 inches were recorded in the extreme northern districts, southern Onitsha and northern Owerri Provinces. In most parts of Enugu, Ogoja, Ahoada, Degema, Port Harcourt, Annang and Calabar Provinces, rainfall values were between 12 inches to 16 inches. Higher rainfall totals for the month of 16 inches to 40 inches were recorded along the coast and also along the border between Eastern Nigeria and the Cameroon Republic.

TABLE 8.6

Climatic data, June 1965

Station	Mean Temp. (°F)			R.H.	Rainfall			
	Daily Mean	Max.	Min.	(%)	Inches	Dept.fr. norm	Rainy days	Dept.fr. norm
Enugu	79	87	72	79	14·2	+7·1	19	+3
Port Harcourt	79	85	72	86	12·5	+1·7	20	−2

JULY 1965

Actual Conditions. The position of the I.T.D. remained north of Nigeria throughout the month, fluctuating between 17° N and 21° N. The pure monsoon rains which had hitherto been limited to the southern parts of the country gradually moved northwards until, by mid-July, all of Nigeria was under the influence of the monsoon air.

The upper air situation was dominated by a series of vortexes (air in a whirling or circular motion) which moved in quick succession from east to west across the Eastern Provinces. A series of line squalls affected the region in consequence.

TABLE 8.7

Climatic data, July 1965

Station	Mean Temp (°F)			R.H.	Rainfall			
	Daily Mean	Max.	Min.	(%)	Inches	Dept.fr. norm	Rainy days	Dept.fr. norm
Enugu	77	85	72	85	9·9	+2·2	21	+5
Port Harcourt	77	81	72	92	20·8	+8·0	24	+3

AUGUST

General Conditions. The occasional dry days which occur at the height of the wet season tend to become sufficiently frequent in August to create what is commonly referred to as the 'little dry season' or 'August break'. In southern Nigeria such a break is quite marked in the west but is far less pronounced in the east (Fig. 8.1).

The 'little dry season' is generally considered to be the result of large-scale atmospheric changes over Southern Africa and the Southern Atlantic (subsiding air masses) which yield a relatively more stable south-westerly air stream over West Africa. This thesis implies that the 'little dry season' is merely a northward extension of the main dry season in the Southern Hemisphere, and that fluctuations or pulsations in the southern anti-cyclone are responsible for the alternation of wet and dry days, also the more continuous three- to four-week run of drier weather, during the Nigerian rainy season.

This plausible explanation is not sufficient in itself to account for the notable diminution of the 'August break' from Western to Eastern Nigeria. The extent to which an air stream can influence weather depends not only on internal characteristics derived from its source region but the effect of local factors, particularly upon its lower layers. The passage of the monsoon winds over the surface waters of the Gulf of Guinea may affect its stability. It is probable that, due to surface ocean currents and the configuration of the shoreline (especially the Niger Delta bulge southwards), the waters of the Bight of Benin have a lower surface temperature than those of the Bight of Biafra, south of Eastern Nigeria. No reliable observations or oceanographic recordings are available to confirm this hypothesis.[1]

[1] A. W. Ireland, 'The Little Dry Season of Southern Nigeria', *The Nigerian Geographical Journal*, v (1962), p. 16.

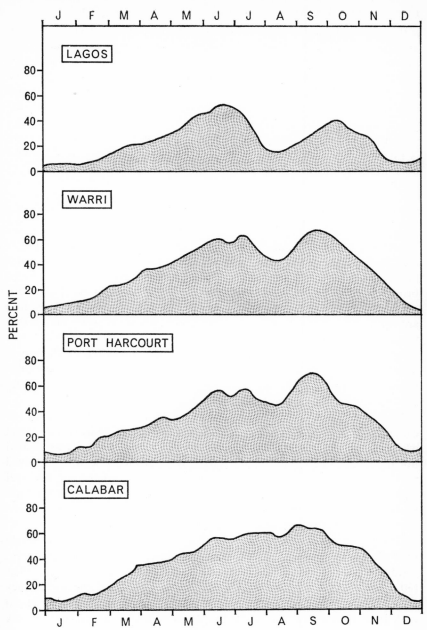

FIG. 8.1 *Percentage occurrence of rainy days (+0·1″) from west to east in Southern Nigeria (after Ireland)*

If this situation pertains, atmospheric instability is greater over the Eastern areas, and rain is more likely to result, thus reducing or even eliminating the 'little dry season' over the Eastern Provinces.

While this mild fluctuation in the pattern of rainfall may appear insignificant in the overall climatic picture, it is of considerable importance to eastern farmers if the 'August break' does occur and – in the exceptional year – persist for two to three weeks. Crop yields may well suffer in consequence and in those areas ill-equipped with water-supply facilities, there is the anomalous situation whereby villagers may have to walk several miles in search of water in the middle of the rainy season.

AUGUST 1965

Actual Conditions. The 'little dry season' which had already begun to set in towards the end of July, especially in Western Nigeria and the Lagos area, began its advance further eastwards. At the beginning of the second half of the month the rainy weather had again returned to almost all areas, bringing to an end the bright spell which had prevailed in most southern provinces. The 'August break' was rather short and even in the West and around Lagos, where its effect was most pronounced, it lasted only about three weeks.

From a position 20° N at the beginning of the month the I.T.D. moved north to about 24° N early in the second week. By the middle of the month it had returned to 17° N, the most southerly position during August. By the end of the month it was back at about 22° N.

During the first week of the month, a strong anticyclonic tendency became rather pronounced over the sea south of Nigeria. During the second week (eighth to fourteenth) this upper-air anticyclone with its associated north to north-west flow inland gradually slackened due to short-lived vortexes over Northern Nigeria. From the twelfth the Saharan sub-tropical high extended into the country, altering the zonal air flow into a light northerly flow. It remained generally wet in the Eastern Provinces, however, with early afternoon L.T.s over high ground.

During the second half of the month an active easterly wave developed over the Cameroon Highlands on the seventeenth and, by the third week, a ridge of the Saharan high extended into Nigeria giving strong easterlies (30 m.p.h.). These winds tended to dampen the effect of the waves they created; widespread squally thunderstorms were nevertheless experienced and, apart from a dry spell on the twenty-third, there was scarcely any day without rain for the rest of the month in Eastern Nigeria.

TABLE 8.8
Climatic data, August 1965

Station	Mean Temp. (°F)			R.H. (%)	Rainfall			
	Daily Mean	Max.	Min.		Inches	Dept.fr. norm	Rainy days	Dept.fr. norm
Enugu	77	84	72	84	13·2	+6·6	19	+3
Port Harcourt	77	81	72	88	21·1	+11·4	28	+6

SEPTEMBER–OCTOBER

General Conditions. The month of September usually witnesses the full-scale resumption of heavy rains emanating from deepening mT air, sufficiently thick to ensure the generation of rain from within itself through 'convergence-producing mechanisms'. Except for the coastal towns of Eastern Nigeria, September is in fact the wettest month of the year.

Following this peak flourish of the wet season, the weather conditions already outlined for the first half of the year are experienced in reverse order. The I.T.D. moves south to a position just north of Nigeria (16° N). The north-easterlies reassert themselves and penetrate further southwards; by late October they are over those areas experiencing humid conditions and thunderstorms in May to June. The steady rains give way to convectional showers and the sharp downpours of D.L. storms with the squall-line belt.

SEPTEMBER 1965

Actual Conditions. A feature of the weather during this month was a series of easterly waves originating beyond the Cameroons which moved rapidly across the Eastern Provinces giving large day-to-day variations in the weather.

During the first week, the I.T.D. moved from 22° N to 20° N. Widespread rains and drizzle and some thundery activity were experienced over Southern Nigeria. In the second week, the I.T.D. retreated southwards to 17° N. Rain showers and isolated thunderstorms continued, the latter becoming more widespread by mid-September. On the fifteenth and sixteenth, a well-defined easterly wave passed across Eastern Nigeria, the moist air was about 4,000 feet deep, and the whole system produced almost continuous rain with embedded thunderstorms.

Towards the end of the third week, the I.T.D. was located at 16° N. Upper winds were mainly easterlies except for occasional vortexes which moved in to disturb the flow. Line squalls with gusts of up to 50 knots in Northern Nigeria produced side-effects in the East: rain and strong winds. At the end of the month (on the twenty-eighth), an upper air wave from the Cameroons produced a cyclonic circulation near the coast with an associated line squall. Very heavy rain resulted in scattered localities.

TABLE 8.9

Climatic data, September 1965

Station	Mean Temp. (°F)			R.H. (%)	Rainfall			
	Daily Mean	Max.	Min.		Inches	Dept.fr. norm	Rainy days	Dept.fr. norm
Enugu	78	87	71	81	11·0	−1·3	19	0
Port Harcourt	78	84	72	85	8·7	−9·0	14	−11

OCTOBER 1965

Actual Conditions. October was in general much drier than average except in a few isolated places. The weather of the month was characterised by late afternoon thunderstorms.

The I.T.D. at the beginning of the month was at 18° N; it moved steadily south until by the end of the second week its position was around 16° N. The migration continued to below 10° N by the end of the month.

On the fifth a well-defined line squall moved into the region from the east and affected the weather during the night and the following morning, giving gusts of up to 35 m.p.h. at several stations. On the tenth, an upper air trough of low pressure along the eastern border moved rapidly across country giving widespread thunderstorms. During the third week, a weak high-pressure ridge developed in the upper air and there was limited convectional activity over high ground in the Eastern Provinces. A number of squally thunderstorms developed in the afternoons but were not pronounced. For the rest of the month, the high-pressure ridge from the Saharan High dominated the region with dry clear weather. Thunderstorms were mostly confined to the coast.

TABLE 8.10

Climatic data, October 1965

Station	Mean Temp. (°F)			R.H. (%)	Rainfall			
	Daily Mean	Max.	Min.		Inches	Dept.fr. norm	Rainy days	Dept.fr. norm
Enugu	80	88	72	78	4·0	−6·2	13	−4
Port Harcourt	80	82	72	87	9·5	−0·6	21	0

NOVEMBER–DECEMBER

General Conditions. The wet season comes to an end with a progressive reduction in the number of convectional and D.L. storms and the reocurrence of rainless yet humid weather. Then the advancing Harmattan re-establishes its authority over Eastern Nigeria, and the dessicating but stable influence of the tropical easterlies is experienced over most districts, particularly in the northern and central parts of the region. The position of the I.T.D. moves firmly from around 15° N to its lowest latitudinal position of around 5° N, i.e. across the southern part of Eastern Nigeria.

NOVEMBER 1965

Actual Conditions. The drier weather conditions which started early in October became more pronounced during November and rainfall was well below normal in virtually all areas of Eastern Nigeria. A premature Harmattan haze was experienced towards the end of the month, with very thick dust reducing visibility to a few hundred yards in places. This was a rather unusual phenomenon for this time of the year, particularly in the coastal districts.

The position of the I.T.D. steadily retreated from 10° N to 6° N by the end of the month. During the first half of November, the winds of the upper air were mainly easterlies and moderately strong. Only a few thunderstorms were reported, mainly in the evenings. The zonal easterlies remained relatively strong (35 m.p.h.) during the second half of the month. The surface weather was marked by extensive Harmattan haze. From 22 to 30 November dust haze spread to extreme southern stations giving periods of intensive dustiness during the day, with fog or mist patches early in the morning along the coast.

One beneficial feature of the Harmattan haze is the extensive deposition of the superfine dust or mineral particles upon the land. Depending on the duration and severity of the haze, several hundred tons of inorganic nutrients may be added to the soils of Eastern Nigeria each year through the agency of the Harmattan.

TABLE 8.11

Climatic data, November 1965

| Station | Mean Temp. (°F) | | | R.H. | Rainfall | | | |
	Daily Mean	Max.	Min.	(%)	Inches	Dept.fr. norm	Rainy days	Dept.fr. norm
Enugu	83	91	70	73	0·0	−2·1	0	−4
Port Harcourt	78	88	72	84	1·1	−2·5	7	−4

DECEMBER 1965

Actual Conditions. The position of the I.T.D. fluctuated between 7° N and 5°N in Eastern Nigeria during December. Dry conditions together with widespread dust haze were still the dominant feature of the weather during the period. Light to variable zonal north-easterlies prevailed in the upper air, interrupted only by a trough of lower pressure which moved slowly westwards from the Cameroon Republic between the ninth and the twelfth. Cool nights and sunny dry days with limited visibility were the common experience of most districts. The only rain received was along the coast where convectional storms gave Port Harcourt a fall of 3.5 inches, a more than 200 per cent departure from the norm for December.

TABLE 8.12

Climatic data, December, 1965

Station	Mean Temp. (°F)			R.H.	Rainfall			
	Daily Mean	Max.	Min.	(%)	Inches	Dept.fr. norm	Rainy days	Dept.fr. norm
Enugu	81	92	67	61	0·2	− 0·7	1	0
Port Harcourt	80	87	70	85	3·5	+ 2·5	2	− 2

SUMMARY OF 1965 WEATHER

The sample year of 1965 was, in fact, an atypical year so far as averages of weather elements are concerned. Mean maximum temperatures were consistently above average for all months, although minimum temperatures were near normal. The wet season began rather later than usual and terminated more abruptly, with a remarkably swift transition from rainstorms to Harmattan haze in November. It was also a much wetter year than normal, as the following table indicates.

TABLE 8.13

Cumulative rainfall for 1965, compared with means for previous years

Station	(1) Cum. Total Jan.–Dec.	(2) Mean Cum. Total	(3) Departure (1)–(2)	(4) % Departure	(5) No. of years of average
Enugu	73·18″	65·53″	+ 7·65″	+ 12%	9
Port Harcourt	104·62″	89·72″	+ 14·90″	+ 17%	15

GENERAL SUMMARY

Some summary observations may now be made on the principal elements of weather and climate in Eastern Nigeria, with particular reference to their import to man in the region.

(i) INSOLATION

Due to its latitudinal location, Eastern Nigeria experiences a high intensity of solar radiation throughout the year. Specific attempts to measure insolation in Nigeria have thus far been few.[1] Daylight hours are nearly constant from month to month; as far north as Nsukka, the difference between the shortest and longest day is only forty-eight minutes. This ensures a steady input of solar energy although absorption, scattering and reflection due to the earth's atmosphere may modify the actual amounts of solar radiation received at ground level. Cloud cover and Harmattan haze can cause an appreciable decrease in insolation rates. Again at Nsukka, the number of bright sunshine hours expressed as a percentage of the maximum possible was only 27 per cent in September 1963 (the rainiest and cloudiest month) as compared with 64 per cent in December 1963 (the driest month).[2]

Nevertheless the overall annual experience of solar radiation throughout the Eastern Provinces is that of marked and sustained intensity. House design, clothing, and social customs – such as the afternoon rest – all reflect the impact of insolation upon the life of the people.

(ii) ATMOSPHERIC TEMPERATURES

In view of the foregoing comment on insolation, temperatures within Eastern Nigeria are predictably high the year round. Mean annual temperatures are everywhere above 75° F although they do not exceed 85° F (Fig. 8.2). Mean minimum temperatures do not fall below 65° F while mean maximum temperatures do not exceed 90° F. These continually high and mildly modulating temperatures, together with the general mugginess created by high relative humidities, are responsible for the unfavourable epithets applied to the Eastern Nigerian climate by expatriates from the intermediate latitudes.

[1] See J. Davies, 'Estimation of Insolation for West Africa', *Quarterly Journal, Royal Meteorological Society*, xci (1965), pp. 359–63.
[2] University of Nigeria, Faculty of Agriculture, *Basic Data: Agrometeorologicall Station, Nsukka* (Nsukka, 1964).

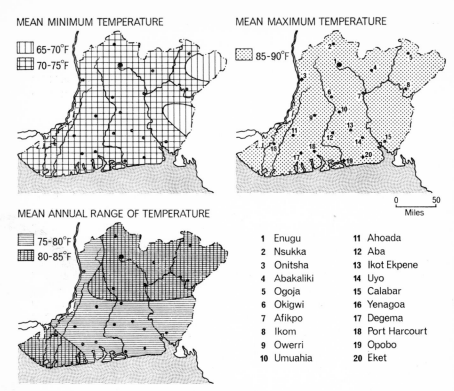

MEAN MINIMUM TEMPERATURE

☐ 65-70°F
☐ 70-75°F

MEAN MAXIMUM TEMPERATURE

☐ 85-90°F

MEAN ANNUAL RANGE OF TEMPERATURE

☐ 75-80°F
☐ 80-85°F

0 ____ 50
Miles

1	Enugu	11	Ahoada
2	Nsukka	12	Aba
3	Onitsha	13	Ikot Ekpene
4	Abakaliki	14	Uyo
5	Ogoja	15	Calabar
6	Okigwi	16	Yenagoa
7	Afikpo	17	Degema
8	Ikom	18	Port Harcourt
9	Owerri	19	Opobo
10	Umuahia	20	Eket

FIG. 8.2 *Mean minimum, mean maximum and mean annual temperatures*
(1951–60)

It is the daily and seasonal march of temperatures rather than their annual means which are more significant to man. Diurnal thermal variations may well be greater than seasonal differences, illustrating the well-known adage 'Night is the winter of the tropics'. Thus Nsukka experiences a mean *daily* range as high as 18.8° F in February, with a mean maximum of 90.8° F and a mean minimum of 72° F, considerably greater than the mean *annual* range of 7.6° F. The marked daily fluctuation in sensible temperatures, i.e. the temperatures experienced by the human body, can cause considerable discomfort to the indigenous population, weakening aged people and infants with colds or pneumonia, and lowering their resistance to infection and disease. Together with frequent soakings in the rainy season, leading to colds and respiratory infections, these weather changes can lead to detectable seasonal increases in the mortality rate.

The hottest months of the year are February and March, and the coldest month is usually August (Fig. 8.3). February and March are late dry season months when insolation is increasing with the apparent northward migration of the sun, the Harmattan haze has largely disappeared, and clearer skies permit uninterrupted sunshine. August is in the midst of the rainy season and the heavy rains of the preceding months have led to a lowering of atmospheric temperatures. Despite the 'little dry season' a high degree of cloudiness persists which deflects incoming solar radiation.

The differences in the range of temperatures at the four stations plotted in Fig. 8.3 reflect the moderating influence of maritime locations, e.g. land and sea breezes (in the case of Port Harcourt and Calabar) also an interior location (Enugu, 745 feet above sea level) and an elevated interior location (Nsukka, 1,300 feet above sea level).

The spatial movement of temperature belts is best indicated in the month-by-month distribution of mean maximum temperatures which clearly illustrates the north–south fluctuation of temperatures in accordance with the interplay of tropical air masses and air streams described earlier.

Temperatures across the region are high enough to permit year-round growth of domesticated plants and agricultural activities. A seasonal pattern to cropping is necessary due to lack of moisture, not a temperature-induced dormant season. Where water can be supplied by irrigation, and the terrain and soils are favourable, year-round agriculture is feasible and may become necessary in the light of the rapid population increase.

Amelioration of temperatures due to elevation occurs along the Nsukka-Udi Plateau, as shown in the bar graphs for Enugu and Nsukka (Fig. 8.3). Together with localised breezes and winds over the upland areas, these climatic factors would reinforce the tendency for populations to settle in the watershed areas for defensive reasons. Only on the Obudu Plateau and in the Oban hills is the altitude sufficient to cause an appreciable drop in temperature. At the Obudu Cattle Ranch the nights are so cool that fires are required and the visitor to the recreational E.N.D.C. hotel sleeps under several blankets.

(iii) RELATIVE HUMIDITY

This is one of the most variable of the climatic elements. In the Harmattan season in particular changes in the relative water-vapour content of the air (associated with diurnal fluctuations in temperature) can be dramatic, as indicated in Table 8.14.

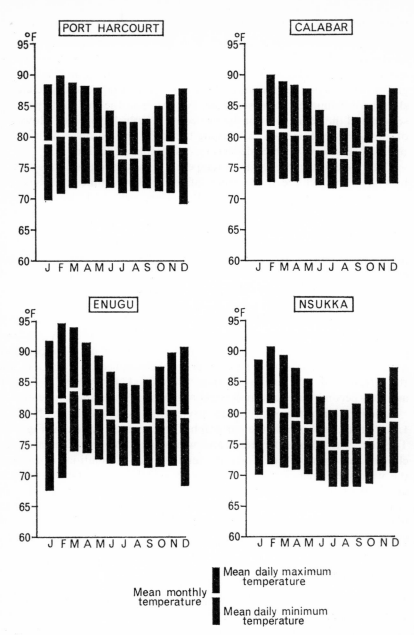

FIG. 8.3   *Seasonal march of temperatures at selected stations*

TABLE 8.14

Variations of R.H. at Nsukka on 8 January 1964[1]

Hour	06.00	08.00	09.00	10.00	11.00	12.00	13.00
R.H. (%)	87	74	52	47	37	36	32

Hour	14.00	15.00	16.00	17.00	18.00	20.00	24.00
R.H. (%)	30	30	32	38	54	57	68

The hourly changes in R.H. at Nsukka for different months of the year is indicated in Fig. 8.5, which also confirms the high degree of variability in this atmospheric phenomenon.

In terms of monthly and annual averages, the R.H. throughout Eastern Nigeria is high, between 70 per cent and 80 per cent. In general it is higher in the vicinity of the coast, as is to be expected, and decreases inland (Table 8.15).

TABLE 8.15

Mean monthly R.H.(%): 10.00 hours, 1965

Station	J	F	M	A	M	J	J	A	S	O	N	D	Annual
Port Harcourt	88	84	80	83	82	86	92	88	85	87	84	85	85
Enugu	77	71	70	73	76	79	85	84	81	78	73	61	76

Relative humidity, as indicated earlier, strongly influences the reaction of the human body to air temperatures. Our ability to maintain constant body heat through perspiring is hampered when the R.H. is high. A marked decrease in the relative proportions of atmospheric vapour may aid the heat regulatory system of the body but excessively dry air can bring other inconveniences such as chapped faces and cracked lips. The alternative name for the Harmattan and its English translation from Hausa – the Doctor Wind – is really only appropriate in those areas, e.g. the coastal lands, where high humidity throughout most of the year is the common experience. But in these areas the Harmattan is experienced, at best, for only a few days out of any one year.

The sudden changes in R.H. over short periods of time also have a marked effect on the efficient functioning of organisms; further studies in animal and plant physiology are required to measure the precise nature of this stress and its consequences.

[1] Ibid.

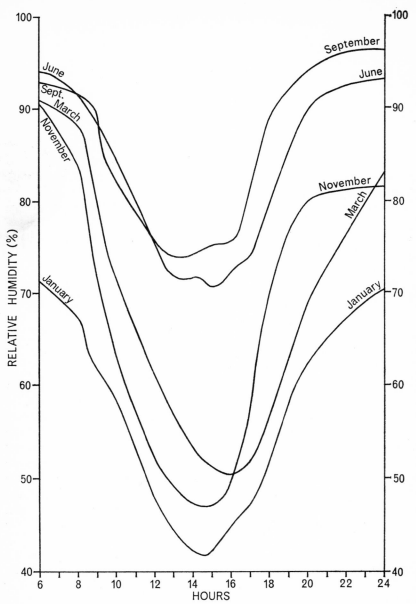

FIG. 8.4 *Mean hourly relative humidity for selected months at Nsukka*
(after Sircar)

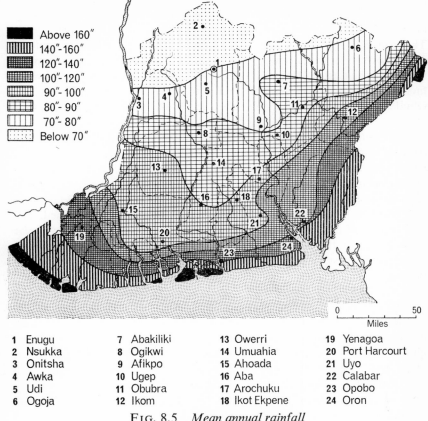

	Above 160″		140″-160″		120″-140″		100″-120″

Above 160″
140″-160″
120″-140″
100″-120″
90″-100″
80″- 90″
70″- 80″
Below 70″

1	Enugu	7	Abakiliki	13	Owerri	19	Yenagoa
2	Nsukka	8	Ogikwi	14	Umuahia	20	Port Harcourt
3	Onitsha	9	Afikpo	15	Ahoada	21	Uyo
4	Awka	10	Ugep	16	Aba	22	Calabar
5	Udi	11	Obubra	17	Arochuku	23	Opobo
6	Ogoja	12	Ikom	18	Ikot Ekpene	24	Oron

FIG. 8.5   *Mean annual rainfall*

(iv) PRECIPITATION

It is the spatial and temporal variations in rainfall which are the most critical component of Eastern Nigeria's weather and climate.

The pattern of annual precipitation is shown in Fig. 8.5. The map indicates that the amounts of rainfall decrease continuously from the coast inland, also from the elevations of the Eastern Highlands westwards. The strong influence of a maritime location and altitude are clearly revealed by the consistent trend of the isohyets. The configuration of the Eastern Nigerian shore is such that the region faces almost squarely the onshore moisture-bearing winds of the Atlantic. In the Delta area in particular, the position of the coast at right angles to the prevailing monsoon winds of the wet season creates an annual rainfall in excess of 160 inches. To the east, the Bonny peninsula receives over 170 inches a year. This convectionally-induced rainfall diminishes rapidly inland at a rate of 1 to 2 inches per mile. Some 30 to 40 miles inland the decrease is much slower. In the Eastern Uplands, orographic precipitation on the windward-facing slopes produces annual figures to rival the coastal figures. The Obudu Ranch received 205 inches in 1962, probably the region's record.

The Central Provinces, 75 to 100 miles inland, have an annual rainfall (80 to 90 inches) which is little more than half that of the coastal locations, while near the northern boundary total precipitation for the year is reduced to approximately 65 inches. Local variations in annual rainfall (due to terrain, vegetational cover, etc.) are disguised at the regional level but may well be significant to the inhabitants of the areas concerned. The Udi section of the Plateau probably receives over 80 inches of rain a year while Enugu (in the rain shadow of the north–south escarpment in relation to the south-west air stream) receives an annual average of just over 70 inches.

In the light of the mean annual figures, all of Eastern Nigeria may be said to experience a tropical wet climate (Köppen's Af climate), which implies a rainfall surplus to the normal needs of the population. However, it is the seasonal distribution of rain which is critical, rather than the overall receipts, for an excess of rain at one season may be followed by a deficit at another. This is the pattern which obtains in Eastern Nigeria, as we have learned. Furthermore, the true measure of the humidity or the aridity of a place is achieved by studying the relationship between incoming rainfall and outgoing water through evapo-transpiration (related in turn to temperature). If rainfall is greater than potential evapo-transpiration or water need, the place is humid. If the reverse is true, the place is arid. In these terms, many parts of Eastern Nigeria are arid for some part of the year, humid for the remaining months. Unfortunately measurements of evapo-transpiration are not yet available over a wide enough area to permit precise calculation of the water balance on a regional basis.

The seasonal pattern of rainfall was emphasised in the month-by-month account of Eastern Nigerian weather, and it is this pattern which so strongly influences the activities of the rural population, and therefore the bulk of the people in the region. The crop calendar for clearing the land, the planting and the harvesting of crops, is a direct reflection of the rainfall regime (Part III, Chapter 10). The timing and intensity of the dry season, the wet season, and the 'little dry season' may be appreciated by examining

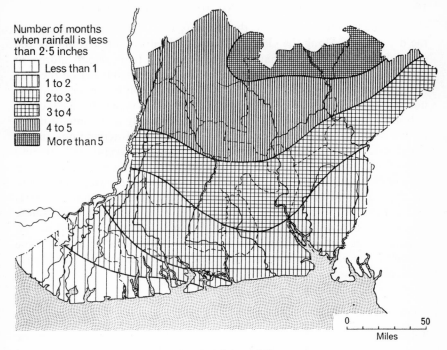

Number of months
when rainfall is less
than 2·5 inches

	Less than 1
	1 to 2
	2 to 3
	3 to 4
	4 to 5
	More than 5

0   50

Miles

FIG. 8.6   *Duration of dry season*

bar graphs of mean monthly rainfall figures for recording stations in
Eastern Nigeria.

The south to north decrease in rainfall is now reflected on a monthly
basis. The severe contrast between wet and dry season receipts of rain is
clearly demonstrated. The bimodal regime of the rainfall (two peak periods)
is discernible at the following stations: slightly in the case of Calabar, Aba,
Afikpo and Abakaliki (generally to the east of the region); moderately in
the case of Degema, Port Harcourt, Umudike (Umuahia) and Okigwi;
strongly in the case of Brass, Bonny, Enugu and Nsukka.

The dry season – taken as months with rainfall lower than 2·5 inches – is
less than one month at the coast, e.g. Bonny and Brass, and increases
gradually inland (Fig. 8.6). In southern inland locations it is two to three
months, e.g. Ahoada and Ikot Ekpene; in central inland locations it is
three to four months, e.g. Umuahia, Owerri and Afikpo; in northern inland areas it is four to five months or over, e.g. Nsukka and Ogoja.

The number of rainy days – days with more than 0·01 inch of rain –

shows a distribution similar to the quantity of rainfall and the length of the dry season. Along the Gulf of Biafra, rain falls on approximately 180 days of the year (approximately 50 per cent) so that schoolchildren in the coastal town of Bonny have an even chance of getting wet on the way to and from school each day. In Nsukka Division, on the other hand, the number of rainy days drops to about 90 a year (approximately 25 per cent).

Finally, deviations from the average daily, monthly and annual rainfall figures may be of crucial importance to farmers. Interruptions in the anticipated pattern, prolonged raininess or drought, exceptionally intense rain over short periods, these and other fluctuations can bring about crop failures through waterlogging or water deficiencies, accelerated soil erosion (as at Agulu-Nanka in violent storms), damage or destruction of homes and communications through flooding, and even loss of life. Absolute maxima of 9 inches of rain in a single day in June (inland locations) and 8 inches a day near the coast in September have been reported. These extremes of weather experience, particularly of precipitation, are clearly much more significant than the modes or means.

(v) WATER RESOURCES

'Of all the exploitable mineral products of the earth, water – or more specifically, groundwater – is the only one which is replenishible.'[1] The prime source of this replenishment is precipitation upon the earth's surface, so that all the water needs of man, animals and plants in Eastern Nigeria are ultimately met from rainfall occurring within the region.

Of the rain which falls over Eastern Nigeria, only a relatively small percentage is destined to reach the water-table, thence to act as a sub-surface reservoir for meeting human needs. Run-off, evaporation and transpiration prevent the rain from penetrating deep into the ground. Run-off is conditioned by slopes, geological formations, soil structure in the surface horizons, also plant cover. Evaporation results from solar radiation; transpiration is related to vegetational growth. The combined action of all these factors in the Eastern Nigerian *milieu* is to produce a high rate of natural withdrawal of ground water and to leave only minor quantities for artificial abstraction (by wells and boreholes) for human consumption.

From an estimated total of 42 billion (42×10^{12}) gallons of rain falling on Eastern Nigeria every year, only some 2·6 billion gallons remain after

[1] R. C. Mitchell-Thomé, 'Average Annual Natural Recharge Estimates of Groundwater in Eastern Nigeria, with Comments on Consumption' (Nsukka, 1963), manuscript, p. 1.

deductions accruing from losses via run-off, evapo-transpiration, etc.[1] Of these 2·6 billion gallons, only vadose or gravity water attains the zone of saturation, perhaps 1·56 billion gallons. But not all of this is destined for human use; for technical, scientific and economic reasons, only about 780,000 million gallons represent recoverable ground water.

The high rate of natural withdrawal understandably fluctuates from place to place within the Eastern Provinces, hence also the modest rate of annual recharging of ground water. Fig. 8.7 shows the areal distribution of hydrologic provinces in rank order. The area of highest potential as regards recharge (and the lowest rate of natural withdrawal) extends south-eastwards from Onitsha to Oron and includes such large urban centres of population as Awka, Orlu, Owerri, Umuahia, Aba, Ikot Ekpene, Uyo and Eket. It also includes the eastern side of the Nsukka-Okigwi Plateau with such population centres as Enugu-Ezike, Nsukka, Udi and Okigwi. The combined area of these two regions of highest potentialities totals some 7,100 square miles or about 24 per cent of Eastern Nigeria.[2]

The second-ranking area totals some 1,400 square miles; together with the first-ranking province, some 29 per cent of the region may be classed as potentially favourable as regards recharge. The most negative areas hydrologically total only 650 square miles or about 2 per cent of the region.

The first- and second-ranking hydrologic provinces include those areas of highest population density in the Eastern Provinces (Fig. 2.2). Thus there is a high positive correlation between settlement patterns and available ground water resources in Eastern Nigeria. If not a contributory factor to the early evolution of the distributional pattern of population, since the ground water supplies were often too deep to reach through shallow pits or primitive wells, the availability of water through modern means of extraction (principally boreholes and pumps) has at least meant that the increasing numbers of people in the high-density zones could be sustained through adequate supplies, some hundreds of feet under the ground.

The question remains: is the annual supply of water to the ground adequate to meet the ever-increasing needs of Eastern Nigeria with its burgeoning population, agricultural expansion and industrial development?

The raising of standards of education and living, economic progress and amelioration of health conditions always involves increased water consumption by populations. A good example of this may be seen in Nsukka Division. In the rural areas, consumption at present varies be-

[1] Ibid., p. 9.   [2] Ibid., p. 8.

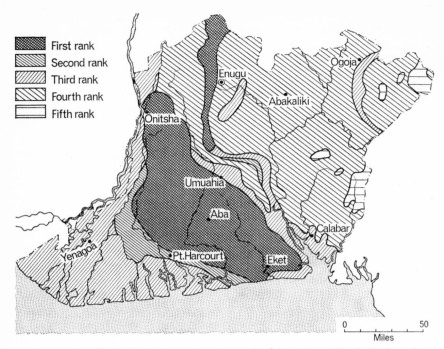

Key:
- First rank
- Second rank
- Third rank
- Fourth rank
- Fifth rank

(Map labels: Enugu, Ogoja, Abakaliki, Onitsha, Umuahia, Aba, Calabar, Yenagoa, Pt.Harcourt, Eket)

0   Miles   50

FIG. 8.7 *Hydrological provinces in rank order (after Mitchell-Thomé)*

tween two to three gallons per day per capita; in Nsukka Town, consumption is about thirteen gallons per day per capita; on the campus of the University of Nigeria, Nsukka, consumption rises to sixty gallons per day per capita.

Estimating the water needs of the future is notoriously difficult because of so many variables: population increase, the shift from rural to urban societies, advances in agriculture (particularly irrigation), industrialisation, and so forth.

Assuming an Eastern Nigerian population of some 14 million in the near future, with a similar rural-urban disposition as at present, Mitchell-Thomé has estimated that the annual water need will be some 77,340 million gallons, a figure well below the 780,000 million gallons which was the estimate of annual recharge.[1] However, the water consumption estimates for 14 million people do not take into account potential industrial, commercial and agricultural needs, which can be extraordinarily high

[1] Ibid., p. 11.

compared with domestic requirements but are virtually impossible to predict. Heavy industries use enormous quantities of water; e.g., to make a ton of pig-iron requires 57 tons of water; a ton of good-grade paper requires 250 tons of water; a ton of nitrate fertiliser requires 600 tons of water. Some of the water may be recoverable and re-usable after industrial use but at considerable expense. In agriculture, irrigation is a greedy consumer of water; great quantities are lost in conveyance through evaporation. It may take up to 4,000 tons of water to raise a ton of rice. This prodigious use of water should be borne in mind when large-scale irrigation schemes are mooted as the answer to increasing agricultural output.

In sum, if we assume the materialisation of all economic development projects visualised under the Six Year Plan, the demand for water may well reach over 300,000 million gallons a year or close on 40 per cent of the annual rate of replenishment. It is conceivable that, by the end of the century, the entire 780,000 million gallons recharge will be needed for domestic, agricultural and industrial purposes. It is imperative then that overall plans be drawn up and implemented for using ground water as economically and beneficially as possible in Eastern Nigeria.[1] Yearly the problems of obtaining sufficient supplies in satisfactory qualities will involve greater effort and financial outlay. We cannot afford to be prodigal with water, Nature's most bounteous gift.

[1] A step in the right direction was taken in 1961, with the commissioning of an Israeli Water Planning Company (Tahal Ltd.) to make a survey of the resources and needs of Eastern Nigeria, and to submit broad recommendations for water development. The survey resulted in the publication of an invaluable volume for those wishing to grasp the scope and problems of hydrology and water resource development in the region. Govt. of Eastern Nigeria, *Water Supply*. The report is supported by ten maps in colour of the region, at a scale of 1 : 1,000,000.

9 Vegetation

INTRODUCTION

THE original or natural vegetation of Eastern Nigeria, as elsewhere in Africa and the world, results from the simultaneous action of a number of factors: climatic (particularly precipitation, relative humidity and temperatures), topographic, edaphic (soils and soil water) and biotic (animals, including man, and plants themselves). This complex of interacting phenomena is frequently referred to as the 'biotic complex' or 'ecosystem'.

Of the factors listed, those of climate have been the most influential in governing the region's original vegetation cover; the mean annual rainfall, together with the duration and severity of the dry season, have determined whether the climax or end-product vegetation would be *forest* (a plant association dominated by woody species, with grasses virtually absent) or *savanna* (an association where grasses are dominant, although woody growth is still present). Thus the physiognomy or outward appearance of plant communities is in large measure a reflection of climatic differences.

Elsewhere, of course, other physical factors have profoundly affected the natural vegetation – low-lying, waterlogged ground in the coastal areas, for example, special pedological characteristics such as marked hydromorphism of soils formed on impervious shales (Chapter 6), the incidence of forest and grass fires, and so forth.

The agency of man has done much to transform the vegetational cover of the Eastern Provinces; human settlement and agricultural land use entail marked modification, even destruction, of the natural vegetation over wide areas. In the rainforest zone of the southern part of the region, with the very high densities of population already noted, the countryside is honeycombed with villages and their farmlands and one is never far from some form of human habitation. In Owerri, Umuahia, Annang and Uyo Provinces, the widespread cultivation of oil palms has produced a distinctive type of vegetation referred to as 'oil palm bush' which is quite dissimilar from the original rainforest. In the northern areas of the Plateau and Escarpment Zone, lengthy occupancy has brought about man-made, 'derived' savanna of tropical grasslands with few trees.

It is important to emphasise that Eastern Nigeria is not, and probably never was, a land completely covered with vast impenetrable rainforests or 'jungles'. This misconception of plant life in the humid tropics, prevalent in Europe and America for many years, resulted from exaggerated traveller's accounts of the vegetational cover in 'darkest Africa' at a time when the contrasts between low-latitude and mid-latitude plant associations were both novel and exciting to European explorers. In fact, the zone of tropical forest which dominates perhaps half the Eastern Provinces today is far from impenetrable and has, furthermore, physiognomic characteristics not unlike those of forested lands found well beyond the tropics.

Fig. 9.1 illustrates the main vegetational zones of Eastern Nigeria. Within each zone there are, inevitably, numerous variations in the actual vegetation types. Relict outliers of rainforest may be found throughout the areas of derived savanna, particularly along the water courses and around 'juju' shrines and other spots sacred to communities inhabiting the zone. Similarly, restricted areas of derived savanna occur in the more northerly sections of the rainforest. In fact the boundaries between zones cannot be sharply defined due to the intermingling of communities of different ecological regions. As Keay has observed: 'the boundary drawn between the Rain Forest and Derived Savanna Zones is meant to represent a line separating "Forest with patches of Derived Savanna" from "Derived Savanna with patches of Forest". In practice there is nearly always a belt, ten or more miles wide, in which the two types are more or less equally represented.'[1]

I. MANGROVE FOREST AND COASTAL VEGETATION

Along the coast, strand and sandbank vegetation appears, mainly in the form of shrubs, grasses and herbs with a rare stunted bush. These species survive in almost pure sand and a perpetually moist, salt-laden atmosphere.

Where deltaic muds and silts accumulate, in tidal creeks and around brackish lagoons, mangrove swamp forests appear. Mangrove trees are easily identified by their unusual stilt-root systems which raise the trunks above the high-tide mark and, by dividing into innumerable laterally-branching rootlets, support the growth of large trees which would otherwise be unable to flourish in the soft mud and swampy environment. The

[1] R. W. J. Keay, *An Outline of Nigerian Vegetation* (Lagos, 1959), p. 6. (Hereafter referred to as *Nigerian Vegetation*.)

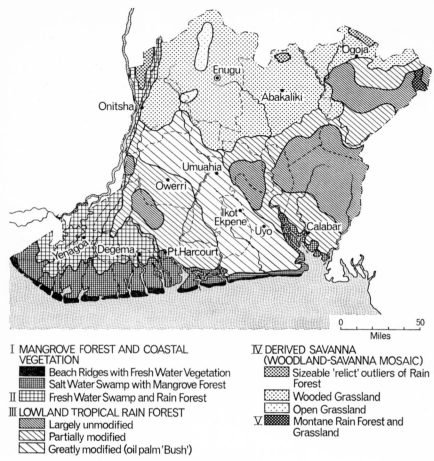

I MANGROVE FOREST AND COASTAL
   VEGETATION
   ▓ Beach Ridges with Fresh Water Vegetation
   ▦ Salt Water Swamp with Mangrove Forest
II ▦ Fresh Water Swamp and Rain Forest
III LOWLAND TROPICAL RAIN FOREST
   ▨ Largely unmodified
   ▨ Partially modified
   ▨ Greatly modified (oil palm 'Bush')

IV DERIVED SAVANNA
   (WOODLAND-SAVANNA MOSAIC)
   ▨ Sizeable 'relict' outliers of Rain
      Forest
   ▨ Wooded Grassland
   ▨ Open Grassland
V ▓ Montane Rain Forest and
      Grassland

FIG. 9.1   *Vegetation*

commonest tree is the Red Mangrove (*Rhizophora*), with an average
height of 30 to 50 feet, although occasional trees 150 feet high and with
an eight-foot girth have been observed.

There are different species of *Rhizophora* with different ecological
requirements so that zonation within mangrove swamps is common and
particularly noticeable from the air. Some forms favour greater salinity
in the soil, others thrive where new deposits of mud are being laid down.
Rhizophora racemosa is the pioneer species and the mangrove tree most
frequently seen in Eastern Nigeria. *R. harrisonii* dominates the middle

areas and rarely exceeds 20 feet in height. *R. mangle* is found on the drier inner margins of mangrove swamp and is more stunted and shrub-like.

The mat of mangrove roots eventually forms a thick, peaty raft which, with the deposition of additional organic materials and detritus, builds up to well above water level and consolidates. The evil-smelling slimy black mud yields to firmer ground and, in turn, the mangrove swamps are replaced by dry-land plant communities. Thus 'the mangrove forest is purely a transitory community continually colonising new ground, stabilising it and then being replaced by other more permanent vegetation'.[1]

Small islands of firmer land within the swamps are frequently used for temporary settlement by Ijaw fishermen in the Niger Delta. In the vicinity of more permanent settlements, mangrove trees are felled for firewood, pit props and building poles. The bark of the Red Mangrove provides *cutch*, a substance used for preserving fishing nets. Where stands of mangrove trees have been depopulated in the less saline areas of the Delta, rice cultivation is being encouraged, with protective bunds to check the influx of salt water resulting from storms in the Bight of Biafra.

II. FRESHWATER SWAMP VEGETATION

Inland from the mangrove swamps, on slightly rising ground and around freshwater creeks and lagoons, entirely different plant communities appear. The land is still low-lying and subject to seasonal flooding but it is not reached by brackish water.

The freshwater swamp forest comprises an irregular growth of many different trees, shrubs, lianas, swamp lilies and grasses. Many of the trees, like those of the mangrove forest, are 'upheld by systems of adventitious curving and ramifying roots (which may be as thick as a man's arm) in such a way that each tree appears perched on a network of intertwined roots'.[2] The canopy or uppermost layer of the forest, created by the merging crowns of the tallest trees, is often one hundred feet high, but it is punctuated by openings caused by changing drainage conditions at ground level. The gaps are frequently choked with shrubs and lianas, and resemble man-made clearings which have been allowed to revert to forest. They are, in fact, naturally caused, for human settlements are rare away from the main water courses.

[1] Ibid., p. 11.
[2] A. M. Aubréville, 'Tropical Africa', in S. Haden-Guest, J. K. Wright, E. M. Teclaff, *A World Geography of Forest Resources* (New York, 1956), chapter 16, p. 363.

Among the trees found in freshwater swamp forests are a number of palms, especially *Raphia spp.*, which provide materials for mats and baskets, also pissava fibre and palm wine. Other useful trees include *Abura* and the mahogany *Khaya ivorensis* for timber, but their exploitation is hampered due to obvious difficulties of transportation.

The prospects for rice production are favourable in this zone and the Ministry of Agriculture has been encouraging community projects for the co-operative clearing and farming of suitable areas, e.g. around Oporama, south-west of Yenaoga. In the fertile triangle created by the divergence of the Nun and Sengana Rivers, the Niger Delta Development Board has erected at Peremobiri a modern rice mill and established experimental *padis* as demonstration units. Unfortunately oil royalties paid to villages in the locality have had a depressant effect upon these agricultural developments in the freshwater swamp zone of the Delta.

III. LOWLAND RAINFOREST (TROPICAL RAINFOREST)

The rainforest zone in Eastern Nigeria, with an average depth today of some eighty miles, formerly extended much further north and embraced the remaining areas in the region which are now classified as Derived Savanna. According to Harrison Church, the northward limit of lowland rainforest may once have been a line joining all places with at least 45 inches annual rainfall, a minimum mean monthly R.H. at 1.00 pm of 40 per cent, and not more than three months with less than an inch of rain.[1] With these criteria as a yardstick, the setting of the regional capital of Enugu, with an average annual rainfall of 70 inches, an average R.H at 3.00 pm of 57 per cent, and only two months with less than one inch of rain, would certainly be clothed in rainforest today were it not for wholesale destruction of the original vegetational cover over many centuries of human occupancy. Similarly Nsukka, with an average annual rainfall of 66 inches, an average R.H. probably close to that of Enugu, and three months with less than one inch of rain, would also be well within the tropical rainforest belt if agricultural exploitation of the area had been less devastating in its impact on the natural vegetation. Within a few miles of Nsukka today, all stages in the degradation of former rainforest can be seen; very old, high, secondary forest survives in escarpment ravines, derived savanna with scrub evergreen and deciduous trees prevails in unfarmed lands around the town, while xerophytic (drought-resistant)

[1] R. J. Harrison Church, *West Africa* (London, 1963), p. 67.

grasses have taken over in the most impoverished areas of red sandy soils. As far north as Enugu-Ezike, near the border with Northern Nigeria, relict patches of high forest are to be witnessed near the centre of the town.

UNMODIFIED RAINFOREST

Man's role in changing the face of Eastern Nigeria has indeed been profound and only in limited areas can anything resembling a mature 'climax' forest still be witnessed (Fig. 9.1). The appearance of unmodified rainforest is misleading to the inexperienced eye. When viewed from an aircraft flying several thousand feet above its canopy, or observed horizontally from a car travelling along a paved road, the forest may give a superficial impression of homogeneity and uniformity: of a continuous expanse of rich, dark-green foliage and straight-trunked, broad-leafed evergreen trees in close proximity. However, a walk through the fecund vegetation soon reveals a heterogeneous collection of trees of all shapes and sizes and in various stages of growth or decay.

The areas of rainforest which have been less persistently penetrated by man tend to have an identifiable structure of layers or strata of plants. The uppermost stratum comprises relatively few species of tall, smooth-barked trees, 120 feet high or more, with wide-spreading crowns which cut down severely on the amount of sunlight reaching the lower layers and the forest floor. Occasional trees rise to 180 feet, even 200 feet in favoured locations. These tall trees of the first layer are appropriately known as emergents and they represent some of the most important timber trees in Nigeria's forests. Among the best known are the 'mahoganies', a commercial name for a number of species, e.g. *Khaya ivorensis*, *K. anthotheca* and *K. grandifoliola*, Obeche (*Triplochiton scleroxylon*), Iroko (*Chlorophora excelsa*), Utile (*Entandrophragma utile*) and Afara (*Terminalia superba*). Due to their great height, many rainforest trees have buttresses of wing-like, triangular-shaped projections which rise eight to twelve feet above the ground and also extend outwards for many feet from the base of the trunk. These support the relatively shallow-rooted forest giants but necessitate the erection of special cutting platforms for felling the trees. Buttress or 'tension' wood has potentialities as a source of wood pulp for paper manufacture.

Another botanical feature of rainforest trees, and a paradoxical one, is that, despite the fantastically high rate of growth of many tropical plants, the tall trees are very slow to grow. Their competing, closely-packed crowns cut down the sunlight and a high degree of cloudiness and humidity also reduces the amount of solar radiation received by the trees. Thus, the majority of the most sought-after African hardwoods take one hundred years or more to mature. Sustained cropping of useful timber through replanting is a lengthy process in consequence, extending over several generations. It is not easy to persuade lumbering companies to look that far ahead into the future.

A second stratum of vegetation consists of a great variety of trees from 50 to 120 feet high, with smaller yet converging crowns which form the main canopy of the mature 'climax' rainforest. According to Keay, the tropical forests of Nigeria, while containing a large number of different tree species by European or North American standards, are less rich floristically than similar forests in other areas of Af climate, in South America, South-East Asia and even Central Africa.[1] Thus, while the species of trees in a square mile of Amazon rainforest may number many hundreds, a comparable area in Eastern Nigeria may barely reach three figures.

The third layer or understorey comprises many smaller trees and younger representatives of the giants, struggling upwards towards the life-giving sunlight. This stratum, up to fifty feet high, also consists of woody climbers, epiphytes and parasitical plants which provide further illustrations of the competitiveness among vegetational forms in the tropical rainforest. The dense understorey is the layer which is most conspicuous to an observer on foot, hampering the view of the upper strata and dominating his impressions of the forest as a whole.

A ground-level association of small, single-stemmed shrubs, herbs, mosses and lichens makes passage through the forest not a difficult matter. There is little accumulation of dead and rotting vegetation, due to rapid decomposition of organic matter. An excellent description of the rainforest ecosystem is given in an account by Gerald Durrell of a zoological expedition in the adjacent Cameroon Republic.

> The forest is not the hot, foetid, dangerous place some writers would have you believe; neither is it so thick and tangled as to make it impenetrable. The only place where you get such thick growth is on a deserted native farm, for there the giant trees have been felled, letting the sunlight in, and in consequence the shorter growth has a chance and sprawls and climbs its way all over the clearing, upwards to the sun. . . . As you enter the forest, your eyes used to the glare of the sun, it seems dark and shadowy, and as cool as a butterdish. The light is filtered through a million leaves, and so has a curious aquarium-like quality which makes everything seem unreal. . . . On

[1] Keay, *Nigerian Vegetation*, p. 17.

every side are the huge trees, straddling on their great curling buttress roots, their great smooth trunks towering hundreds of feet above,
their head foliage and branches merged indistinguishably into the
endless green roof of the forest. Between these the floor of the forest
is covered with the young trees, thin tender growths just shaken
free of the cradle of leaf mould, long thin stalks with a handful of
pale green leaves on top. They stand in the everlasting shade of the
parents ready for the great effort of shooting up to the life-giving
sun. In between their thin trunks, rambling across the floor of the
forest, one can see faint paths twisting and turning. These are the
roads of the bush, and are followed by all its inhabitants.[1]

Within Eastern Nigeria, the rainforest to the east and south of the
Cross River has somewhat different characteristics to that over the rest of
the region, since the river appears to have formed the western limit of a
number of plants common in Central or Equatorial Africa but rarely
found beyond the Cross River, either in Western Nigeria or the remaining
forest areas of West Africa. The forests in the Oban hills are, for example,
more completely evergreen than those on the Coastal Sands of the south-
western part of the region. The marked contrast in population densities
has led to further differentiation in the physiognomy and floristic composition of the rainforest in the two areas.

At present, commercial exploitation of Eastern Nigerian rainforest on a
large scale (through foreign investment) is occurring in only one location,
south of the Cross River near Obubra (in south-east Abakaliki Province),
where Brandler and Rylke Ltd. have established a modern sawmill and
are felling in the northern fringe of the Oban forests. By an agreement
with the Government, which calls for restocking of worked-over areas, the
company has access to more than 400 square miles of high forest, with an
option to work an additional 400 square miles at a future date. The sawn
timber is ferried over the Cross River and transported by road to Enugu,
thence by rail to Port Harcourt for export – largely to the U.S.A. (12,000
tons were exported in 1965).

OIL PALM BUSH

Over much of the rainforest zone west of the Cross River few areas of
undisturbed timber remain. Virgin rainforest in other than small, remote
and isolated stretches has virtually disappeared. High secondary-growth
forest is more abundant, particularly north-west of Port Harcourt, but it is

[1] G. Durrell, *The Overloaded Ark* (Penguin Books, 1228; London, 1961), pp. 30–31.

neither as lush nor as luxuriant as the original plant cover. With the penetration of new roads and the emphasis on agricultural development in the form of government-sponsored plantations which require clear- or 'salvage-felling', even the remaining areas of secondary forest – with the exception of the forest reserves, the most extensive of which are east or north-east of the Cross River – are destined to vanish.

In the eyes of foresters and timber conservationists, the chief enemy of their trees has always been the peasant farmer. Traditional methods of indigenous farming which involve clearing the forest by fire and matchet, cultivating the cleared land for a few years, then allowing it to revert to bush fallow (Part III, Chapter 10), have over the years resulted in wholesale destruction and decimation of the tropical rainforest. In its place, again over wide areas of central Eastern Nigeria, the preservation and cultivation of the oil palm (*Elaeis guineensis*) has given rise to a man-induced association of plants or oil palm bush.

The dominant trees are clearly the palms, from 50 to 100 feet high; while not confined to the rainforest belt, they are nevertheless very much at home there after the loftier trees have been removed, and are to be seen in profusion today. In a journey from Enugu to Uyo, via Umuahia, or to Port Harcourt, via Owerri, the traveller is inevitably impressed with the sea of oil palms through which he must pass. Occasionally solid stands of the formerly wild, ragged-leafed trees are seen in the vicinity of villages. Many of the trees are old, and poor producers of the clustered bright-red fruit, but they remind one that the trade in palm oil and kernels is a venerable one in Eastern Nigeria. It was not for its mineral oil or petroleum that Southern Nigeria was formerly known as the 'Oil Rivers Protectorate'. Privately-owned, commercially-operated and government-run plantations are now increasingly in evidence, with improved low-growing NIFOR stock (Part III, Chapter 11).

The palm forests also supply poles and leaves for building and thatching huts, although present-day improvements in house construction call increasingly for mud or cement blocks and corrugated-iron, tin-pan roofing. Many over-tall palms which no longer produce fruits of commercial value are tapped for the ever-popular *tombo*, or palm-wine. A cross-section of degraded rainforest trees and shrubs, even patches of savanna, are found in association with the palms in oil palm bush. Other economic trees are also present, however, particularly those bearing edible fruits such as bananas and plantains, citrus fruits – oranges, limes and grapefruit – pawpaws (papaya), mangoes and avocado pears.

IV. DERIVED SAVANNA

When second-growth rainforest is persistently slashed and burned, and the period of bush fallow is shortened, the soils become increasingly impoverished and the vegetational cover deteriorates to an open woodland of fire-resistant, often scrubby trees, climbers and shrubs. Grasses progressively invade the farm lands of the forest zone, until the original high forest is replaced by derived savanna or woodland-savanna mosaic, with only vestiges of the former plant associations.

The spread of tall savanna grasses, of such species as *Andropogon, Imperata* (Spear grass), *Loudetia* and *Pennisetum purpurem* (Elephant grass), means that the fallow areas are ravaged annually by fires, started both deliberately and by accident. The fires encourage still further the growth of grasses which send up new shoots shortly after the burn, but many of the less-resistant trees are killed. With the mounting invasion of grasses, the fires become still fiercer until only the most fire-tolerant savanna trees remain, species such as *Phllanthus discoideus, Dialium guineense*, the *Combretum* climbers and the oil palm.

Within the area of derived savanna in Eastern Nigeria, shown in Fig. 9.1, there are varying combinations of grasses and trees depending on the intensity of human settlement, soils, slope and drainage. The chief belt of open grassland with only scattered trees (orchard savanna) appears on the Nsukka-Udi Plateau, particularly within the zone of porous, deep-red soils derived from sandy deposits. In the vicinity of Nsukka, or along the main road from Enugu to Awka, near the Oji River, savanna grasses may be seen to dominate the natural landscape.

Elsewhere in the derived savanna zone, in Onitsha and Abakaliki Provinces, for example, savanna woodland is mainly in evidence, on reddish-brown, sandy and gravelly soils derived from sandstones and shales. The density of trees may exceed fifty to the acre, but varies widely according to the degree of human interference. *Eugenia spp., Protea, Uapaca guineensis* and scrubby *Vitex detarium* are characteristic trees; while on the richer, loamy alluvial soils of the Anambra Basin, tall *Daniellia* and fan palms (*Borassus aethiopium*) are dominant.

Outliers of rainforest may still be found across the woodland-savanna mosaic: along rivers such as the Mamu, draining to the Anambra, on ridges occupied by villages in the rolling countryside, within the protected ravines along the Nsukka-Enugu Escarpment, and around village settlements in the upper reaches of the Cross River Basin. Indeed, village bush,

F

comprising preserved species of many rainforest trees, accounts for the most numerous occurrences of relict high forest; areas of religious significance in the early animist and ancestor-worshipping societies have most frequently been left in thick forest cover. Such sacred groves are a clear indication that high tropical forests once extended over the entire northern sector of Eastern Nigeria. As observed earlier, the northernmost supervillage of Enugu-Ezike has patches of rainforest adjacent to its geographical centre. The forest reserves on the border with Northern Nigeria in Ogoja Province and at Effium, north of Abakaliki, also testify to the former extent of rainforest.

V. MONTANE VEGETATION

A marked increase in elevation within the tropics can lead to important modifications of the vegetational cover, due to lower average temperatures, greater rainfall and cloud cover and more constant humidity. Steep slopes and their effect upon the drainage patterns of mountain areas are also significant factors. The montane biotic complexes differ appreciably from those found in lowland locations.

In Eastern Nigeria, a small extent of montane vegetation occurs on the Obudu Plateau, the western extension of the Cameroon Highlands into the region. Here in the Sonkwala Mountains, extending to nearly 6,300 feet, montane rain and mist forest occupies most of the lower slopes and consisted originally of a rich growth of trees, epiphytes, tree ferns and mosses, reflecting the high humidity of the area. Today, the forest vegetation ceases abruptly around 5,000 feet and is replaced by montane grasses which grow to knee-height in compact tussocks. The sudden change is indicative of controls other than the climatic elements of temperature and rainfall. Within the area occupied by the E.N.D.C. cattle ranch, the tree line invariably coincides with a sharp change and intensification of slope, the forest being found in the sheltered, steep-sided river valleys below the general level of the rounded hills which dominate the ranch landscape. In such locations, the forest is protected from the fires which periodically sweep across the exposed grassland areas of the Plateau. The grasses represent, in all probability, the fire-climax vegetation resulting from many centuries of human occupancy of the uplands for defence, food production and the rearing of livestock. The controlled ranges of the Obudu ranch now provide excellent, tsetse-free grazing for sizeable herds of beef cattle.

FOREST RESOURCES

Mounting concern is being expressed, particularly by Forestry Department officials, at the continued rapid depletion of the forest resources of Eastern Nigeria. The unplanned encroachment of migrant farmers, the piecemeal expansion on the part of land-hungry communities, government-sponsored Farm Settlements, Rural Development Projects and various plantation schemes, are together taking a visible and heavy toll of the natural vegetation of the region. Less obvious but no less severe is the annual depredation of the forests and woodlands for firewood. The principal fuel for domestic purposes throughout Eastern Nigeria is wood, and consumption in villages and towns amounts to an estimated 20 million tons a year. While coal, natural gas and lignite are abundant, it is not possible to forecast the time when the country will be so economically advanced as to have electricity and gas in every home.

Yet the forests themselves represent a valuable source of high-quality furniture woods and building timber for both internal and world markets. They may also provide foodstuffs such as nuts, fruits, oils; gums, resins and fibres; and medicinal ingredients. Forests also function as conservators of soil and they protect catchment areas from excessive run-off of rain water. Their ability to induce rain is scientifically more questionable, although the high transpiration rate from innumerable broad leaves must add vast quantities of water vapour to the local atmosphere.

Protection of forests through the establishment of forest reserves was initiated with commendable foresight early in the century, but the areas set aside are not large. Table 9.1 gives a precise measure of the areas involved in Eastern Nigeria.

TABLE 9.1

Forest reserves in Eastern Nigeria, 1962

Area of Forest Reserves (sq. miles)			High Forest as % of total reserves	Forest reserves as % of total area of region
Total	High Forest	Savanna Woodland		
2621·43	2489·00	132·43	94·95	8·9

Several positive measures in forest conservation may be adopted in the years ahead to improve on the present situation, at the same time not to impede the very necessary developments in agriculture and other forms of land use for the economic advancement of Eastern Nigeria.

The enlargement of the public forests through the planting-up of areas unsuited for crops is the first possibility, with particular emphasis upon firewood plantations for meeting the growing domestic needs of the population. At the same time urgent attention should be given to the possibilities of a large-scale production of cheap coal and lignite briquettes so as to lessen the reliance upon wood as a cooking fuel (such a move would also serve to revitalise the distressed coal industry). Next, an accurate resources inventory of existing forest reserves is necessary, taking into full account accessibility, growing stock, net growth and allowable cut, to encourage more efficient utilisation of these resources.

Most important of all, systematic provision for renewal of the forest resources should be made. The government's arrangement with Brandler and Rylke for the replanting of cut-over areas is a model for similar agreements with both large and small contractors in the future. Multiple land use within forested lands should also be encouraged, whereby farmers and commercial exploiters of timber alike may benefit. A co-operative and mutually beneficial system of land use between cultivators and foresters along the lines of the *taungya* system developed by the British in Malaya is worthy of adoption in Eastern Nigeria. Areas from which valuable timber has been removed are farmed along scientific lines for some years, during which time the land is also being regenerated with planted species of commercial trees. Implicit in this scheme is the realisation that reliance upon natural vegetation must give way in the future to dependence upon man-created reserves of exploitable timber; tree-farming should be the ultimate goal in Eastern Nigeria. Also multiple land use, whereby the land is not locked up entirely in forestry enterprises but may also be utilised for cropping or grazing, is a desideratum if not a requirement for the future.

> Basic to all considerations of policy is the fact that Nigeria – a 'forest nation' – is nearing the end of its supply of timber. A shift from dependence on natural forests to those grown under some measure of human control is required in the years that lie ahead. In this new industry, silviculture will not depend on the original vegetation of the country, but on soil and water resources. Thus, as the original wealth disappears, new problems emerge, namely soil and water conservation and irrigation projects.[1]

The satisfaction of future internal requirements of wood, also the world's need of tropical timber, will best be met from plantations of commercially important trees, worked on a sustained-yield basis, with or without an

[1] S. Kolade Adeyoju, 'The Forest Resources of Nigeria', *The Nigerian Geographical Journal*, viii (1965), p. 125.

association with enlightened farming communities. The managerial control required for implementing such a forest policy presents real problems but is an exciting challenge to future land-use planners and administrators in Eastern Nigeria.

PART THREE

Economic Activities in Eastern Nigeria

10 Traditional Agriculture

INTRODUCTION

AGRICULTURE in Eastern Nigeria employs over seventy-five per cent of the working population, utilises a comparable percentage of the land surface of the region (although well over half of this is in bush or grass fallow: (Fig. 10.1)), and contributes – together with the processing of agricultural products – a significant share of the Gross Domestic Product (G.D.P.). Until the discovery and exploitation of oil, agricultural commodities made up some eighty per cent of the region's commercial assets. Over fifty per cent of the region's output still originates in the agricultural sector. Agriculture is therefore a major determinant of the geography of Eastern Nigeria and deserves our close attention in consequence.

The field of agricultural geography covers the various ways in which man uses the land for producing crops and raising livestock (crop and animal husbandry). All plants which yield agricultural products are termed crops, and the 'crop' is considered to be the harvested product, whether it be seed, fruit, roots, forage, wood or other product. The term 'livestock' covers domesticated animals and birds such as cattle, sheep, goats, pigs, chickens and ducks.

The geographer is concerned to identify and analyse the spatial patterns created on the land by farmers and to examine the variable characteristics of agricultural activities over the earth's surface. In co-operation with other scientists, geographers in Eastern Nigeria have many opportunities for contributing to a fuller knowledge of agriculture in the Eastern Provinces. Field mapping and land-use surveys on a topographical scale[1] can assist the presentation of an accurate picture of crop patterns and agricultural methods at the regional level, an essential basis for designing those innovations in agriculture which are so patently necessary for the economic development of Eastern Nigeria. Written analyses of traditional and modern agricultural systems, derived from the geographer's vantage point, may

[1] R. P. Moss, 'Land Use Mapping in Tropical Africa', *The Nigerian Geographical Journal*, iii (1959), pp. 8–17.

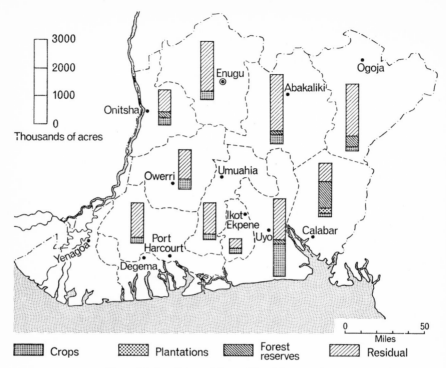

Legend:
- Crops
- Plantations
- Forest reserves
- Residual

Scale:
3000
2000
1000
0
Thousands of acres

0 — 50 Miles

FIG. 10.1   *Disposition of land by provinces*

also make their contribution towards resolving some of the intractable problems which face the rural land-use planner in Eastern Nigeria and the Republic as a whole.

A pioneering geographical survey of agriculture in Nigeria was carried out by K. M. Buchanan in the early 1950s and is reproduced in his excellent (if dated) regional text on the country.[1] The framework for the treatment of agriculture in that study has been adapted for use in the present work.

ESTABLISHED AGRICULTURAL ECONOMIES

The traditional, well-established agricultural economies may be differentiated in a number of ways. Here we shall consider them on the basis of:

[1] K. M. Buchanan and J. C. Pugh, *Land and People in Nigeria*. The Human Geography of Nigeria and its Environmental Background, chapter iii, 'The Agricultural Economy', pp. 100–70.

i. the systems of cultivation utilised;
ii. the crops and livestock produced.[1]

After a description of these aspects of traditional agriculture, an analysis of the contemporary problems arising from the persistence of earlier farming methods is made and suggestions offered as to how these problems may perhaps be overcome.

(i) SYSTEMS OF CULTIVATION

The techniques of farming pursued by traditional subsistence cultivators[2] are the heritage of centuries of adaptation and adjustment by peoples with a common store of attitudes, objectives and technical skills (culture) to the realities of the physical environment. They reflect a hard-won core of empirically-derived knowledge concerning soil and crop capabilities within the familiar *milieu* of the homeland.

> The cultivation system of an area represents an equilibrium established by man, usually after many long centuries of trial and error, between himself, the physical environment (climate and soil) and what may be termed his 'biological auxiliaries' (plants, animals and microfauna), and upon which his survival and progress depend.[3]

Due to the innate conservatism of many farming communities, such time-honoured methods of land use may persist long after the conditions which originally called them into being have disappeared or have drastically altered. Changing attitudes and objectives on the part of younger generations and marked modifications to the natural environment may not therefore reflect in changing patterns of land use until some time after the agents of change have emerged. It is this cultural time-lag which is responsible for certain of the outstanding problems of agricultural land use in the Eastern Provinces today.

Nevertheless, change is of the essence of agricultural systems since, sooner or later, adjustments to meet new situations in the physical or human

[1] It is also possible to analyse agricultural activities in terms of the ultimate consumer of the harvested products; this is a common approach of economists.
[2] The term 'subsistence cultivator' implies that the farmer concerned is supporting himself entirely from his own agricultural efforts. In fact, few areas of Eastern Nigeria have a purely subsistence economy; interdependence with other parts of the region and the country as a whole have led to an *internal exchange economy*, involving cash transactions for the sale of excess crops or the purchase of needed foodstuffs. Elsewhere, agricultural products are produced for the world market, leading to an *external exchange economy*.
[3] Buchanan and Pugh, *Land and People in Nigeria*, pp. 101–2.

geography of an area are inevitable. To guide these changes so as to achieve the most beneficial and harmonious man–land relationships in the new circumstances, with the least amount of disruption of the old socio-economic order – this should be the ultimate goal of all land-use planning in rural areas.

In view of the considerable diversity in both the original and man-modified physical environment in Eastern Nigeria, as described in Part II of this study, also the ethnic diversities of the population covered in Part I, Chapter 1, we may anticipate a broad range of human responses to the necessity of producing foodstuffs. Differences in size and shape of cultivated plots, in crops raised, in the crop calendar (dates for clearing the bush, planting, weeding and harvesting), in agricultural implements, in storage facilities, and in social customs related to the harvesting and consumption of food, all these may be expected to occur from place to place in response to variations in soils, slopes, drainage, terrain, vegetational cover, rainfall, tenurial systems, population size, available labour force, dietary preferences, marketing arrangements and so forth.

With the foregoing *caveat,* it is possible to identify five general systems of cultivation by traditional farmers in the Eastern Provinces:

(*a*) shifting cultivation;
(*b*) bush fallowing;
(*c*) rudimentary sedentary cultivation;
(*d*) intensive sedentary cultivation: 'compound' farming;
(*e*) intensive sedentary cultivation: terrace farming.

(*a*) *Shifting cultivation.* There is an extensive literature on this technique of farming in the tropics, also on the system of bush fallowing (*b*) which is highly similar in its methods.[1]

Shifting cultivation, or migratory agriculture, entails periodic movement of fields or plots of crop land (land rotation), also the relocation of hamlets or villages after the area in the immediate vicinity of the settlement has been worked over for agricultural purposes. Short periods of land occupancy (one to three years) by domesticated plants are followed by long periods of natural bush fallow (ten to fifteen years). The system demands that large areas should be accessible to small populations; the man/land ratio may be in the order of twenty to thirty persons per square mile. For these reasons, the areas where shifting cultivation is still practised in Eastern

---

[1] For one authoritative account, see Pierre de Schlippe, *Shifting Agriculture in Africa: the Zande System of Agriculture* (London, 1956).

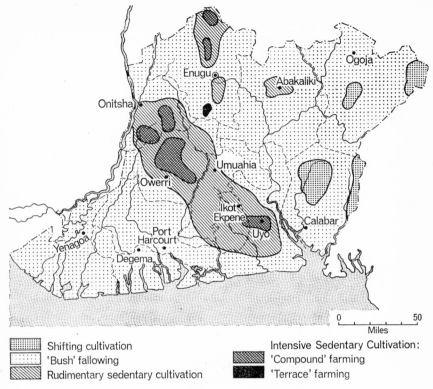

Shifting cultivation
'Bush' fallowing
Rudimentary sedentary cultivation

Intensive Sedentary Cultivation:
'Compound' farming
'Terrace' farming

FIG. 10.2 *Traditional systems of cultivation*

Nigeria are minimal (Fig. 10.2), although it was undoubtedly more wide-spread in earlier centuries.

Small plots an acre or so in size are selected on the basis of favourable soils, revealed through plant indicators, slope and drainage characteristics, and even taste. The natural vegetation (rainforest) is cleared during the dry season through the laborious process of 'ring-barking' or firing the base of the larger trees, cutting down trees of more manageable size with matchets, and rooting out smaller trees and bushes. Only the forest giants are left standing, also economically useful and fire-resistant trees such as the oil palm, oil bean and castor bean trees.

After some weeks, the dead and decaying vegetation is sufficiently dry to permit firing, and a little before the rains are expected, the entire clearing is burned over, leaving an ash layer to provide additional nutrients for the first planting. With the advent of the early rains, the softened soil is then cultivated into small hemispherical mounds by means of short-handled hoes. This patterning of the ground into 'heaps' or mounds tends to concentrate the surface layer of humus and ash for the benefit of seed yam, cassava cuttings and other seeds which are placed in the mounds at planting time, usually in March. They also serve to keep the growing crops above the water level in flat or low-lying areas where waterlogging of clayey (hydromorphic) soils may be experienced.

Apart from an occasional weeding during the wet season and the training of vines around stakes, little further attention is given to the plot until harvest time. This cycle of cropping is repeated for a year or so until diminishing yields indicate depletion of the natural fertility of the soil, after which the plot is abandoned for agriculture and allowed to regenerate under bush or forest fallow. Eventually, under true shifting cultivation conditions – when all the usable land within a reasonable distance of the village has been exploited in the manner described – the community of farmers will migrate to a fresh site, when the whole process of clearing the high forest and engaging in patch farming will be repeated.

The appearance of agricultural landscapes resulting from shifting cultivation and bush fallowing is, to Western eyes, a most untidy one. Small irregular patches of imperfectly-cleared forest contain a mixture of crops and weeds in what seems to be entirely random disorder. The riot of vegetation and tangled greenery of inter-cropped plants such as the tubers (yam, cassava, cocoyam), cucurbits (plants of the gourd family: melon, squash, pumpkin and handle-shaped gourds), maize, beans, etc., presents a well-nigh chaotic scene when compared with the regimented rows of wheat, maize or vegetables on European and North American farms. Yet it is a mistake to assume that old-order, subsistence African farming is in fact as crude or inefficient as its casual and unplanned appearance would suggest. Given the limited objectives and technical skills of traditional cultivators, and the environmental handicaps of a humid tropical environment, this response by Eastern Nigerian farmers to the challenge of growing essential foodstuffs within the rainforest ecosystem was a remarkably effective and wise one. Indeed, it proved an admirable adaptation to the limitations of the physical habitat, so long as – and this is an essential proviso – the number of people engaged in shifting cultivation remained small and abundant land was available for so extensive an exploitation of the natural resources.

The adversities of the natural environment have been considered at length in Part II. Soils are at best moderately fertile, but largely low in plant nutrients and liable to lose fertility rapidly through leaching and oxidation of humus after being cleared of vegetational cover. Heavy rains from squall-line thunderstorms occur, carrying essential minerals down into the subsoil beyond the reach of growing plants, and inducing sheet wash and gully erosion. High solar radiation raises exposed soil temperatures to the point where organic content is rapidly consumed. The techniques of shifting cultivators were designed to deal with these problems in a number of ways. Multiple cropping, with plants arriving at maturity at different times, ensures a continual canopy of vegetation, giving protection to the soil from the climatic elements. Inter-cropping of plants with differing and complementary nutritional requirements ensures that the limited soil fertility is used to the full. Furthermore, it provides protection against total crop failure as may happen in monocultural systems.

> Imperfect clearing of natural cover, and imperfect cultivation of the growing crops, although they may result from the inadequacy of African farm tools, also serve to provide shade and protection to the soil and to assist in re-establishment of bush or forest. The small, isolated fields are more easily recaptured by the natural vegetation than larger ones would be; at the same time spread of plant disease is much slower and more difficult than it is in the vast fields of America or Europe.[1]

In sum, shifting cultivation was a remarkably effective and wise response to the natural setting. Regeneration of the soil could only be accomplished under the long-term restorative effects of nature. Farmers equipped with the simplest of implements could wrest an elemental living from the forest.

But despite these advantages of migratory subsistence agriculture, the system – together with the allied techniques of bush fallowing – clearly has enormous limitations and inefficiencies for communities that are increasing rapidly in size, that have access to finite and often restricted land resources, and are aspiring to standards of living beyond those of largely self-sufficient and economically-retarded societies. Shifting cultivation and other traditional forms of agriculture simply will not sustain the economic progress which is being called for by the mass of underprivileged peoples in Eastern Nigeria and elsewhere in Africa. The monumental challenge of ascertaining how best to transform traditional agriculture into a more pro-ductive force for the betterment of society will be discussed in Chapter 11.

(*b*) *Bush fallowing*. The foregoing description of shifting cultivation is

---

[1] W. O. Jones, 'Food and Agricultural Economies of Tropical Africa', *Food Research Institute Studies*, ii (1961), p. 9.

equally applicable to bush fallowing, except for the periodic uprooting of settlements and their relocation in areas of unfarmed rainforest. As soon as population densities much in excess of thirty persons per square mile developed, tribes began to stake out territorial limits to their domains and unrestricted access to forest lands for extensive farming became more and more difficult. Eventually the fluid pattern of semi-nomadic agricultural societies began to jell and settlements assumed permanent locations within established tribal boundaries. With rising numbers, the prospects of contact and also of conflict between different ethnic groups increased, necessitating the selection of defensible sites and the establishment of permanent protective features such as ditches, pits, walls and – in the high forests – belts or rings of preserved, closely-growing trees.

Thus there began to emerge the patterns of settlement and systems of farming which are still to be witnessed in many parts of the Eastern Provinces today. The food-producing areas of most Eastern Nigerian villages traditionally fall into two geographical categories:

(1) The compound, or inner farmlands (Ibo: *ani uno, ani mbubo*), which are tended usually by the women for raising their kitchen-garden crops, although compound land also contains fruit trees such as bananas, plantains, citrus and kola.

(2) The main or outer farmland (Ibo: *ani agu*), cultivated largely by the men on a bush fallowing basis for producing the staple food crops – yam, cassava, cocoyam, maize, beans and cucurbits. In some areas a third or intermediate zone prevails (often comprising land unsuited for cropping) which is reserved for thatching-grass, used for the perennial renovation of roofs in the settlement.

At present, bush fallowing or field/forest rotation systems for the outer farmland (locally referred to as 'block' farming in Iboland) prevail in those zones where populations range from low to medium and even to high densities; Fig. 10.2 and 2.2 may profitably be compared. Where the natural forest vegetation has, through constant clearing and firing, reverted to a woodland-savanna mosaic or derived savanna, the regenerative processes of grass-fallowing supersede those of woody plant forms.

As is to be expected, a great variety of bush fallow or slash-and-burn techniques exists in response to varying pressures on the land, differences in resource base, and cultural preferences in farming procedures and staple foodstuffs.[1] The duration of the period of cropping, and particularly the

---

[1] For a valuable description by a cultural anthropologist of variations in bush fallowing techniques at eight sample villages in Eastern Nigeria, see D. Smock, *Agricultural*

period of fallowing, may fluctuate considerably; in low-density areas, plots are cultivated for one to three years, then allowed to rest for fifteen years before being reopened for further cropping. At the other extreme, in areas of high densities and acute pressure on agricultural land, fields may be farmed for four years in succession, then rested only for a comparable period of three to four years before being planted up again. It is not possible to map the broad spectrum of farm/fallow ratios in land rotation agriculture at the regional level since, to the writer's knowledge, no systematic survey of the proportional periods of cropping and resting has been attempted.

Needless to say, the beneficial effects of the long-term natural fallow of fifteen years or more (by which time the soils have virtually recovered their natural structure and fertility) are entirely lost when the bush fallow period is shortened to a bare three to four years. The recuperative role of regrowth vegetation has little opportunity to assert itself; similarly the crude efforts of farmers to sustain soil productivity within the cultivated patches, through elementary crop rotations for example, are scarcely sufficient to prevent the mounting drain on soil resources.

At the same time, it is important to note that the rotation of crops (empirically arrived at by native farmers) has been allied with land rotation in those areas of Eastern Nigeria where the increased numbers of people to be supported has meant a marked reduction in the length of the natural fallow period. A typical pattern of crop rotation as practised in parts of Nnewi (Onitsha Province) is shown in Table 10.1. When the land is first cleared of bush, it is planted to the prestige crop, yam, which is intercropped with four or five other plants. In the second year the principal crop becomes cocoyam, while the same inter-crops are planted. During

TABLE 10.1

Modified bush fallow farming: Nnewi (Onitsha Province)

1st year	2nd year	3rd year	4th year	5th–8th years
Yam	Cocoyam	Yam	Early	Late Cassava
Maize	Maize	Maize	Cassava	and Bush Fallow
Beans	Beans	Beans	Beans	
Pepper	Pepper	Other	Other	
Okra	Okra	Vegetables	Vegetables	
Other	Other	Cocoyam		
Vegetables	Vegetables			

Development and Community Plantations in Eastern Nigeria (Lagos, 1965), Pt. I, pp. 1–16. (Hereafter referred to as *Agricultural Development*.)

the third year, yam again assumes pride of place, but by the fourth and fifth years the depleted soil is fit for little more than cassava. After the second cropping of cassava is harvested, some eighteen months after planting, the land reverts to a short-term bush fallow for three to four years before the cycle is repeated. With modifications to suit local conditions and needs, this rotational pattern may be taken as a model for many areas where modified bush fallow farming is practised. In certain areas in addition, as for example on the Nsukka Plateau, a local bush *Acioa barteri* is deliberately planted as a cover crop during the short fallow period. This plant is noted for its soil stabilising qualities and also provides straight sticks which may be utilised to support the yam vines during the next cropping cycle.

Despite these voluntarily-adopted patterns of crop rotation, the progressive deterioration of the land for agricultural purposes is undeniable. It is the steady reduction in the length of the natural fallow which has set into motion across Eastern Nigeria the distributing spiral of diminishing yields of crops and mounting degradation and erosion of soils. Farming communities are often unwilling or economically unable to substitute the beneficial short-term effects of chemical fertiliser mixes for the long-term restorative effects of nature, or to adopt other government-recommended practices for maintaining productivity. In the circumstances, communities still practising the traditional systems of bush fallowing are inevitably faced with further reductions in yields and serious deterioration of the resource base upon which their livelihood depends. Scientific measures which might be adopted to reverse these distressing trends are considered in a later chapter.

(*c*) *Rudimentary sedentary cultivation.* Mounting pressure of population under land-rotation systems eventually reaches a point beyond which it is no longer feasible to have fifty per cent or more of a community's agricultural land lying fallow. Even if yields are lowered, as assuredly they must be if alternative measures for maintaining or restoring fertility are not adopted, it is necessary for as much farmland as possible to be put under cultivation, in an effort to meet the subsistence requirements of the population. It is at this stage that serious cracks in the structure of traditional rural societies in Eastern Nigeria appear. On the one hand, the plots of cultivated land become smaller, with the division of holdings among claimants to land of the next generation; to this fractioning must be added the fragmentation of holdings, and the handicaps which arise when farmers attempt to work a number of dispersed and often widely-separated plots.

On the other hand, a marked exodus of young adult males from the community occurs: landless school-leavers and older men (even women) dissatisfied with conditions in their place of birth. This largely rural to urban migration is also a feature of areas where bush fallowing is still practised but in the present communities it begins to attain more disturbing proportions. When male absenteeism amounts to 50 per cent or more of the adult segment of the society one may seriously question whether the social coherence of extended families or clan groups can be maintained. Only the sustained loyalties of the 'sons abroad' have thus far kept many rural settlements from falling apart.

The impact on traditional agricultural land use of dense populations is therefore one of virtual abandonment of bush fallowing, and the substitution of almost continuous cropping of the outer farmlands, together with a more intensive utilisation of the inner or compound land. The areas of sedentary cultivation in Eastern Nigeria are shown in Fig. 10.2 and coincide with the zone of high to very high population densities.

The voluntary techniques now adopted in an effort to keep the soil in good heart are numerous and commendable as far as they go. The rotation of crops receives serious attention, with a more careful selection and sequential planting of crops in an effort to maintain yields and to husband the soil's productivity. A novel feature is the introduction of non-food-producing cover crops such as the wild *crotolaria* which binds the soil and protects it from the climatic elements during both the wet and the dry seasons. The importance of legumes such as beans and pigeon peas as soil builders is also clearly recognised although, according to professional agronomists, their ability to fix nitrogen under tropical conditions is still open to question. Again near Nnewi, the pattern of crop rotation shown in Table 10.2 is practised.

TABLE 10.2

Rudimentary sedentary cultivation: Nnewi (Onitsha Province)

1st year	2nd year	3rd year
Yam	Maize or Cocoyam	Late Cassava
Beans	Beans	Pigeon Peas or
Vegetables	Vegetables	Cucurbits
	Early Cassava	*Crotolaria* Fallow

This pattern of cropping is pursued in a staggered fashion by many Nnewi farmers on different portions of their holdings, so that yams will always be available for harvesting at each year's 'New Yam Festival' in August.

In addition to encouraging the growth of *crotolaria* in successive dry seasons between cropping, also as a cover crop in the third year's growing season, leaf mulching and dry mulching are practised. The former requires much labour in gathering leaves from the forest areas and spreading them on the land to cut down loss of soil water by evaporation, and eventually to add organic matter to the soil. Dry mulching entails tilling the surface of the soil at intervals during the dry season, again in an effort to reduce evaporation (this 'dry farming' technique is commonly practised in the semi-arid Great Plains of North America by commercial grain farmers). To check sheet and gully erosion of soils in the wet season, cultivators in the high-density, sedentary farming areas frequently tie their heaps or mounds together by connecting ridges, or throw up continuous ridges of earth along the contour. These may be linked by cross-ties which encourage surface water to soak into the sandy soils rather than running away between the lines of ridges (this technique resembles 'basin-listing' which is also carried out in the drier lands of North America as a water conservation measure).

Finally, composting with vegetative organic matter and manuring with animal organic matter are increasingly undertaken, along the lines of practices developed more intensively within the inner farmlands. Rudimentary sedentary cultivation in the outer crop lands is thus partaking more and more of those techniques for intensive production of foodstuffs which have been evolved on compound land. In addition to the cultivation of annual root, cereal, legume and gourd crops, the wisdom of leaving land under permanent tree cover has come to be appreciated so that sizeable acreages of economic tree crops such as the oil palm, banana, plantain and citrus are a further feature of areas of sedentary cultivation.

Impressive as these adaptations by traditional farmers are to the environmental and social realities of the present day, and supporting as they do densities of population far higher than any other rural areas in sub-Saharan Africa, it is regrettably a fact that most communities are fighting a losing battle in their efforts to maintain soil productivity and to feed the ever-increasing numbers of people emerging within their boundaries. Other, more radical transformations in agricultural land use in the high-density areas of Eastern Nigeria are necessary, if a way of life other than 'the meanness and meagreness of mere existence' is desired for the peasant farmer and his family.

(*d*) *Intensive sedentary agriculture: compound farming*. In the original design, the compound gardens adjacent to the dwelling units were utilised for producing vegetables and fruits which were required daily by the womenfolk in their preparation of meals. The diminutive plots of land were carefully tilled and tended, and made the beneficiary of enriching treatments of domestic refuse and trash swept from the homes and compound yards. This material frequently contained animal organic matter in the form of goat or fowl droppings, also human wastes. A small-scale but intensive system of continuous cropping resulted, resembling in its family labour inputs the patterns of allotment gardening in the U.K. and 'victory' gardens in the U.S.A.

With the expansion of hamlets and villages in response to the growth of population, the areas of compound-type gardening increased proportionately. The rising demand for staple foods, and diminishing yields from the outer farmland due to the shortened periods of bush fallow, led to the raising of basic starchy foodstuffs (yam, cocoyam) in addition to vegetables within the compound plots, also an extension of the intensive techniques of composting and annual cropping from the compounds to the main farming areas immediately adjacent to the village. Thus rudimentary sedentary systems of land use were born out of the experiences and accumulated skills of compound farming.

The compound-type system of intensive farming is found within the heart of virtually all Eastern Nigerian villages today, while its adoption in the peripheral areas of settlements on former outer farmland is evident in the very high density areas extending from Onitsha to Uyo, and again on the Nsukka Plateau (Fig. 10.2). Among the large and super-villages of Orlu, Okigwi and Nsukka Divisions, one observes amidst the groves of oil palms small plots scarcely 0·1 and 0·2 of an acre, devoted to gardens of yam, cocoyam, cassava and maize. The palm bush areas are studded with these pocket-handkerchief patches of crop land, and with the mud-walled compounds and numerous habitations of the local population. Within the compound walls are further diminutive plots of garden vegetables. The impression gained is one of a multi-tiered system of farming with several levels of crops being maintained on the same few square yards of soil: an ingenious technique of mixed and sequential cropping. Lowly ground-creeping plants such as the cucurbits are shaded by the foliage of root crops, which in turn are found growing within the shade of the smaller 'economic' trees such as the banana, plantain, orange, mango, breadfruit, native pear, cashew, castor bean and kola; the ubiquitous oil palms grow taller, varying in height from some thirty to sixty feet, and are only eclipsed by scattered specimens of forest giants from the original rainforest, especially cottonwood trees. In sum, as many as a dozen food-yielding, beneficial plants may be struggling for survival and fruition in hard proximity.

This system is maintained by the regular application of compost and manure. Goats and sheep are kept in enclosures and fed grass, leaves and kitchen wastes. The droppings of the animals are compounded with the decomposing litter. Poultry are similarly confined in sheds overnight and their droppings carefully gathered for addition to the compost pit of decaying vegetation which is a common sight in most compounds.

Baskets full of compost and manure are carried to the fields prior to planting. Placed in the furrows between the rows of soil ridges from the previous season, the compost is buried in the process of constructing new ridges on top of the old furrows. Seed and top dressings of organic matter are also made at planting time. It is in this laborious and painstaking manner that rural densities of 1,000 persons or more per square mile are in part sustained by local agriculture. But the system in itself is incapable of supporting the entire population. Imports of foodstuffs are necessary paid for by the proceeds of trading and, more important, the remittances from 'sons abroad'. The rate of male absenteeism in the areas of compound-type farming may be as high as 60 to 70 per cent. The system is thus entirely the affair of 'those at home': that segment of the population largely older men and women, which has not left home to seek employment elsewhere. This hard core of elderly farmers, and the widespread distaste for the tedious routine of gardening on the part of young men, makes rural education for the reforms which must sooner or later inevitably be introduced a formidable task.

(e) *Intensive sedentary agriculture: terrace farming.*[1] In Eastern Nigeria intensive terrace-type farming on steep slopes was formerly widespread on the Nsukka Plateau, in response to the need to produce foodstuffs in the environs immediately surrounding the fortified villages on the Upper Coal Measure outlier hills. Most of the hilltop sites are now abandoned, the people having moved down to the undulating sandlands of the Plateau proper with the advent of more peaceful conditions. Following the exodus from the hills, the terraced slopes were also abandoned; today, the 'fossil' forms of early terracing are a striking sight on the grassy, treeless slopes of many hillsides around Nsukka. Only in limited areas – near the villages of Obimo and Aku for example – do the hills still support small oil-palm

[1] For a fuller treatment, see B. N. Floyd, 'Terrace Agriculture in Eastern Nigeria: the Case of Maku', *Nigerian Geographical Journal*, vii (1964), pp. 91–108.

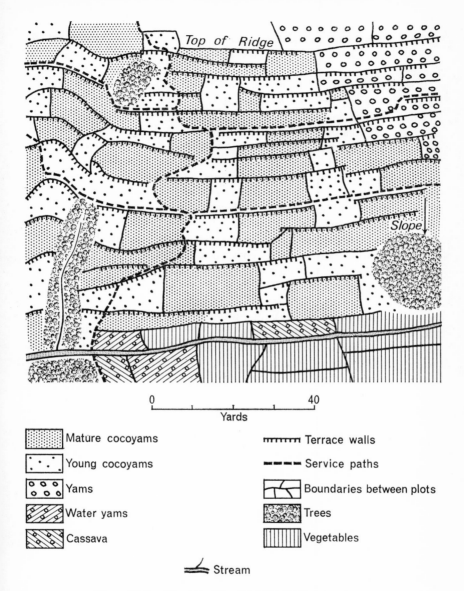

Top of Ridge

Slope

Scale:
0 — 40
Yards

Legend:
- Mature cocoyams
- Young cocoyams
- Yams
- Water yams
- Cassava
- Terrace walls
- Service paths
- Boundaries between plots
- Trees
- Vegetables
- Stream

FIG. 10.3   *Sketch-map of Maku land-use pattern*

villages on the summits and the terrace platforms are still cultivated, albeit on a rotational basis, with some plots resting under grass fallow conditions for several years.

It is further south, in the area known as Maku, near Awgu (thirty miles south of Enugu), that the most conspicuous and impressive illustration of extant terraces may be witnessed in Eastern Nigeria today (Fig. 10.2). On all sides, the hills of Maku present a checkerboard-like mosaic of minute fields situated on numerous terraces and filled with growing crops: yam, cocoyam, maize and vegetables. The terrace walls, varying from one to five feet in height, are constructed of fragments of ferruginised sandstone. The terrace platforms are from five to ten yards wide. The land on the terraces remains gently sloping: completely level terracing has not been achieved although the tendency over the years is for the cultivated plots to become more nearly horizontal. Neither are the terraces continuous, that is they do not form a continually-running bench around a hillside but are interrupted and reformed at irregular intervals.

The carefully constructed terrace walls are aligned roughly along the contour and, although interrupted, they tend to overlap those above and below so that run-off cannot gather momentum on its downward path and the soil is protected in consequence (Fig. 10.3). Within the plots of arable land between the retaining walls, particularly on the steeper slopes, the heaps in which root crops are placed also run generally along the contour and are inter-connected to create virtually a continuous contour bank or bund. Clearly the utmost care has gone into the physical conservation of every square yard of soil in the cultivated areas.

A systematic rotation of crops is practised by Maku farmers, the selection and sequence of crops in rotation showing considerable variation both in space and time, indicating a well-developed sense of the potential of different elevations, exposures, slopes and soils for the production of different crops. A remarkable feature of agricultural organisation among the Maku is that all farmers owning plots in a particular location tend to follow the rotation of crops best suited to that area. This produces a homogeneity of domesticated plant cover; one hill at a certain season of the year may be largely under cocoyam; another may consist of plots filled with yam or yam with its various intercrops. A high degree of self-imposed discipline is thus evident on the agricultural landscape. Each farmer works in unison with his neighbour and, in consequence, a group-spirit pervades the utilisation of the soil. For those farmers with five or six plots of land, often widely separated on different hillsides or in different valleys, this implies a working association with several groups of landowners and considerable variety in farming activities from place to place and season to season.

In sum, the present cultural landscape in the Maku region gives an impression of harmonious and fruitful adjustment by man as farmer to his hilly *milieu*. But in reality the soil is under a continual and relentless pressure to produce foodstuffs for a population which is outgrowing its means of survival. Despite the superficial appearance of an efficient and prosperous agricultural community, statistics of production and consumption gathered during a rural economic survey of Maku record an increasingly frustrating struggle to win subsistence from an austere environment with but few natural resources, and these already sorely over-tapped. There seems to be no alternative to the fact that a greatly increased rate of emigration from the terraced hill land must be achieved if those who remain are to have any sort of life other than one of constant drudgery and privation.

(ii) CROP AND LIVESTOCK PRODUCTION

An indication of the principal crops being grown by traditional farmers in Eastern Nigeria has been given in the preceding section. We may now turn to consider some of the regional variations in the patterns of production which in turn reflect differences in the physical and human geography of the Eastern Provinces.[1]

An important feature of Eastern Nigerian agriculture is that food crops grown for internal consumption claim over seventy-five per cent by value of the total agricultural production of the region. With the exceptionally high population level for a territory of this size in tropical Africa, this emphasis on subsistence food crops is understandable although it means that the greater share of the sweat and labour of rural residents goes on meeting daily dietary requirements, and fails to contribute to the earning of essential foreign capital for economic development through the export of cash crops.

A second significant aspect of agriculture in the Eastern Provinces is the dominant role of yams. Of all the food crops grown in the region, the yam is the only domesticated plant which is indigenous to West Africa (Guinea corn or sorghum occupies only very minor areas in the northern districts of Eastern Nigeria). Since it has been in cultivation much longer than any

[1] This section draws upon manuscript material supplied by L. Uzozie, of the Geography Department, University of Nigeria: *Patterns of Crop Combination, Concentration and Intensity of Production in Eastern Nigeria.*

other staple food crop in the rainforest zone, the study of yam is likely to reveal more about the environmental and cultural factors affecting farming in Eastern Nigeria than the other, imported crops. At present, yam accounts for over fifty-seven per cent by value of the total food-crop production.

Yams are rarely grown alone but occur in mixture with other basic food crops. The identification of crop-combination regions is a first step towards assessing the importance of yam from place to place in Eastern Nigeria. The number of times that yam appears in combination, and the percentage area of the total crop land it occupies within any defined crop-combination region, can be used as an index of its significance in traditional agricultural systems in the Eastern Provinces.

The standard food crops produced in quantity in Eastern Nigeria are the root crops of yam, cassava, cocoyam, the grain crop of maize, the pulse crop of pigeon pea, and the tree (sucker) crop of plantain. Other food or speciality crops produced in smaller quantities are beans, cucurbits, groundnuts, bambara nuts, rice, millets, guinea corn, benniseed, sweet potatoes, sugar cane. The oil palm, an export crop discussed in Chapter 12, also provides oil and kernels for domestic consumption.

If only the six primary food crops are selected, it is possible to identify nine major crop-combination regions:

(a) imperfectly-developed monoculture based on yams;
(b) imperfectly-developed monoculture based on cassava;
(c) yam / maize;
(d) cassava / yam;
(e) yam / cassava / maize;
(f) plantain / cocoyam / cassava;
(g) cocoyam / plantain / cassava / yam;
(h) cassava / yam / cocoyam / maize;
(i) yam / cassava / cocoyam / maize / pigeon pea.

By means of statistical processing of data drawn from some 155 villages surveyed in sample censuses of agriculture between 1959 and 1964, Uzozie was able to produce a map of crop-combination regions (Fig. 10.4). The relative importance of each crop may be measured in terms of the number of times they are drawn into the combinations. Yams and cassava struggle for the first place with seven each across the region. Both cocoyam and maize appear in four combinations. Plantain is drawn twice but in both instances in the wet south-western districts while pigeon pea appears once, in the dry north-west.

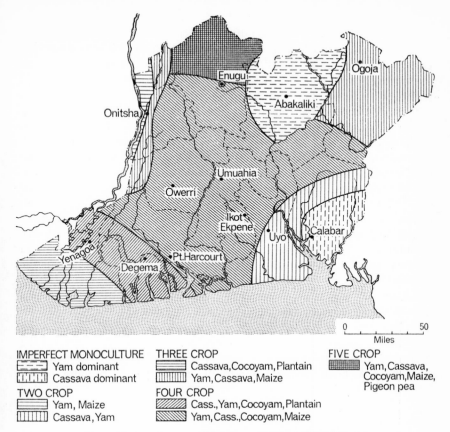

IMPERFECT MONOCULTURE
▭ Yam dominant
▥ Cassava dominant

TWO CROP
▤ Yam, Maize
▥ Cassava, Yam

THREE CROP
▤ Cassava, Cocoyam, Plantain
▥ Yam, Cassava, Maize

FOUR CROP
▨ Cass., Yam, Cocoyam, Plantain
▧ Yam, Cass., Cocoyam, Maize

FIVE CROP
▦ Yam, Cassava, Cocoyam, Maize, Pigeon pea

FIG. 10.4   *Crop combination regions (after Uzozie)*

In analysing the patterns of the crop-combination map, Fig. 10.4 may profitably be compared with Fig. 10.2, of systems of traditional farming, Fig. 2.2, of population distribution, Fig. 6.1, of soils, and Fig. 8.5, of rainfall. There is a general correlation between four and five crop regions and the zones of medium to high population densities, also rudimentary sedentary and compound systems of farming. The desire to produce as much food as possible in the high density areas while insuring against crop failures was commented upon above. However, areas of very light densities also coincide in part with marked multiple cropping (southern Ogoja, western Port Harcourt, Provinces). In these areas, shifting cultivation or simple bush fallowing is still practised, with the chaotic rather than the orderly admixture of plants described previously.

Areas of imperfect monoculture coincide with the lightly-settled hydromorphic (gravelly and clay) soils of Abakaliki Province and the equally sparsely-settled ferralitic soils derived from granites and sands in southern Calabar Province. Cassava is reputed to be more tolerant of adverse conditions than yam, being less demanding of soils, rainfall and topographic conditions. Cassava is shown growing all the way from the mangrove swamps of the Niger Delta to the Oban Uplands: from the high rainfall Coastal Sands Plain to the drier, northern Nsukka Plateau. Yet, with the exception of the saline soils of the Delta, yams appear within an equally wide spectrum of environmental conditions. In fact, in the perennially flooded areas of the Niger Valley, yam with its shorter life cycle has a clear advantage over cassava, which takes from twelve to eighteen months to reach maturity and cannot therefore safely be planted in areas liable to inundation. Maize is a quick-growing but 'greedy feeder' crop, and hence shows up at its strongest in the two-crop area of the Niger-Anambra alluvial flood plain. Cereals and pulses become more important in the northern sectors with their lower rainfall.

Among the speciality subsistence crops, the cucurbits are the most widely distributed due to the ease with which they can be interplanted with the major tubers of yam and cassava, forming a useful cover crop in addition. Rice is associated with the fresh-water swamps of the Niger and the Anambra, and the hydromorphic soils of the Cross River Basin in Ogoja and Abakaliki Divisions. This important cereal has great promise for the future as a crop capable of high and sustained yields on soils of limited fertility, hence the bulk production of starchy foodstuffs for the ever-expanding populace. The other grains – millets, benniseed, Guinea corn – are limited to the drier northern sections, as are the leguminous crops – beans, groundnuts, bambara nuts, pigeon pea – all of which have migrated to the region from their natural environment in Northern Nigeria.

A more precise picture of the areal distribution and concentration of yam production, is provided by the dot distribution map, Fig. 10.5. However, it should be noted that the acreages devoted to any particular crop give only a partial measure of that crop's importance, since land planted to a high-yielding crop may be reduced to make room for a lower-yielding crop which is nevertheless in demand.

The highest concentration of yam appears again in Abakaliki Province, where some 4,000 square miles are dominated by the crop to form a 'Yam Triangle'. Emphasis on yam is particularly noticeable on imperfectly-drained soils derived from Asu River Shale which are less suited for other traditionally-respected crops such as maize or cocoyam. Until the introduc-

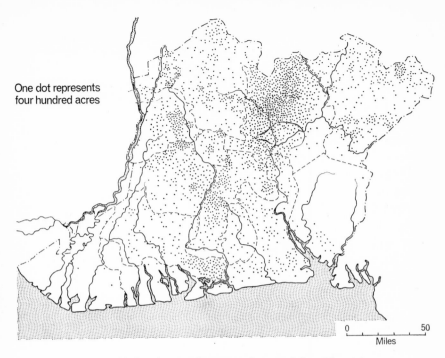

One dot represents
four hundred acres

0   50
Miles

FIG. 10.5   *Areas of yam production (after Uzozie)*

tion of rice in the early 1940s yam held undisputed pride of place in the
agricultural endeavours of the Abakaliki Ibos, whose limited numbers
gave them access to ample land for surplus production of this important
cash crop. Initially the overcrowded Ibo and Ibibio areas within the high
density axis, with declining yields due to reduced fallow periods, provided
an outlet for the surplus yams. With improvements in transportation, yams
from Abakaliki began to find national markets; this encouraged the farmers
to further their specialisation despite the labour-intensive measures
necessary (i.e. the construction of massive heaps or mounds) to avoid
damage to the tubers from waterlogging during the rainy season.

From the Yam Triangle in Abakaliki the concentration falls sharply in
most directions, particularly to the south-east where, due to a sparse
population and inaccessibility, yam acreages are negligible. The overcrowded parts of Iboland and Ibibioland, with heavily leached soils,
shorter fallow periods and a major alternative cash crop (the oil palm),
devote only modest acreages to yam.

The major areas of cassava production coincide with the high density zones of Owerri and Umuahia and are a clear reflection of depleted soils and lower yields per acre. Processed cassava or *garri* is not as popular as yam as a basic foodstuff, but the ability of cassava to grow on soils which will fail to produce a satisfactory yam harvest must be recognised as a significant factor in explaining its distribution.

A surprisingly large acreage of cassava appears in Abakaliki despite the emphasis on yam production in that province. The established demands on the Abakaliki area for staple crops in the food-deficit rural areas and the towns of the region has encouraged the commercial production of cassava and rice in addition to yam. The influence of terrain and low population reflect in the small areas of cassava in the Niger Delta and the Eastern Uplands.

Tree Crops. Significant contributions to the diet of Eastern Nigerians are derived from tree crops as well as from roots and cereals; tree crops have also traditionally provided an important source of income for peasant farmers on smallholdings.

The principal subsistence forest crops are fruit trees such as bananas, plantains, citrus (orange, grapefruit), pawpaw (papaya), avocado (alligator pear), mangoes; also kola, oil bean, castor bean, coconut palms and – most valuable of all – the oil palm, which has also served over the years as the main export crop from Eastern Nigeria, based almost exclusively on collection of the fruit from wild or semi-wild trees by indigenous farmers. Newer tree crops of mounting importance to the agricultural economy of Eastern Nigeria, introduced during the first half of the twentieth century, include rubber, cocoa, cashew and coffee. These are considered further in Chapter 12, since they are largely produced under plantation systems of agriculture.

As we have learned, the traditionally important trees are found growing on the compound lands and in the environs of all rural communities where, especially in the areas of dense settlement, the natural vegetation has yielded to thick groves of oil palm and other economic trees. No official survey has been held to measure the production and internal consumption of tree crops but it is obvious that many thousands of tons of healthful fruits are consumed yearly. The oil palm is of outstanding importance to the external exchange economy of Eastern Nigeria. Nigeria as a whole is the single largest supplier of palm produce in the world, responsible for about 38 per cent of the palm oil and kernels entering into the world market In 1963 Eastern Nigeria supplied, via the Eastern Nigerian Marketing

Board, 95 per cent of Nigeria's palm oil exports (120,000 tons) and 50 per cent of the kernel exports (200,000 tons), producing a revenue of £19 million for the Marketing Board. A further 60,000 tons of palm oil were consumed within the region. Virtually all of this impressive output (95 per cent) comes from the oil palm bush of the village lands, although plantation production is rapidly increasing.

It is difficult therefore to exaggerate the importance of the oil palm to the rural residents of the region. Used in the cooking of starchy foods derived from the tubers, palm oil is a valuable source of vitamins in the indigenous diet, while palm wine obtained by tapping the tree provides a universally popular intoxicating drink. But, even more important, it is an essential source of revenue for subsistence cultivators, unable to produce sufficient basic foodstuffs to maintain themselves and their families throughout the year. The degree of reliance upon this tree is a reliable index of the breakdown of traditional crop farming in the high density zones.

The clusters of red palm fruit are gathered largely from wild trees preserved by farmers during bush fallowing operations over the years, and now grouped together in self-generating groves. Many of the trees are old and tall, fifty to sixty feet in height, which makes collection of the fruit a hazardous operation; the density of the trees may vary from ten to a hundred per acre. In low-density areas, an excessive amount of time is spent in gathering the fruit and transporting it to processing centres. In areas where the trees are crowded together, yields of fruit and oil per acre are exceptionally low.

Simple but wasteful hand and foot maceration of the fruit for extraction of the oil accounts at present for some forty per cent of the product. About 4,000 hand screw presses yield a further fifty per cent, while 118 Pioneer Oil Mills with more elaborate machinery account for the remaining ten per cent. During the five months of the year when harvesting and processing of the fruit are undertaken (approximately from December to April) evidence of the operation abounds. Piles of the bunched fruit may be seen by the roadside where bush trails from the interior farmlands intersect main communication routes. Lines of women with head loads may be seen along the roadside, converging on the presses in the villages. Heavily-laden lorries bring clusters of fruit from village producers' societies to the Pioneer Oil Mills (established by the Eastern Nigeria Development Corporation, although some are now in the hands of private entrepreneurs), strategically located at river crossings where water is available for processing the fruit in the dry season. The smoke and steam from wood-fired boilers, rising above the roofs of the mills, are further indications of the hive of activity associated with the handling of this significant crop, whose expressed oil and kernels are shortly to proceed many thousands of miles for conversion to soap, cooking fats, lubricating greases and many other products in the industrial towns of Britain and Western Europe.

The greatest constraint in the palm oil industry stems from the poor quality and low yields of the semi-wild trees at source, and wasteful processing techniques. In recent years the old groves of unimproved trees have proved an increasingly unsatisfactory source of supply for competing with the rising output of superior-quality oil palm products from commercial plantations in other West and Central African states, and from Indonesia in South-East Asia. An indication of the falling quality and productivity of oil palms in Eastern Nigeria is given in Table 10.3, which records oil palm production from all sources over the period 1958–63. The trend revealed is that of a stagnant or declining overall output, and a steady drop in the output (though not the proportion) of plantation and special grade palm oil. Only in 1963 was the downward trend in output arrested and a start made on returning to the record production figures of 1960, the year of independence.

TABLE 10.3

Palm oil production in Eastern Nigeria, 1958–63[1]

(tons)

Year	Plantation and Special Grade	Technical (Lower Grade)	Total
1958	146,223 (84·1%)	27,659 (15·9%)	173,882 (100%)
1959	144,393 (80·8%)	33,944 (19·2%)	178,337 (100%)
1960	142,660 (79·6%)	36,568 (20·4%)	179,246 (100%)
1961	140,595 (82·1%)	30,713 (17·9%)	171,308 (100%)
1962	113,515 (85·8%)	18,725 (14·2%)	132,240 (100%)
1963	130,988 (85·8%)	21,529 (14·2%)	152,517 (100%)

It is intended to increase the percentage of oil extracted by promoting the use of Stork Hand Hydraulic Presses in Eastern Nigeria, also improving the efficiency and capacity of the Pioneer Oil Mills. Upgrading the quality and yield of promoting trees is being tackled by a tree-crop improvement programme, involving incentive payments and technical assistance to co-operating farmers. A Palm Grove Rehabilitation Scheme (part of a broader Tree Crop Subsidy Programme covering rubber and cocoa also) was initiated in 1962, to continue to 1968, at which time it was

[1] Eastern Nigeria, Ministry of Agriculture, *Annual Report, Agriculture Division 1963–64* (Enugu, 1966), Table 6, p. 15. (Hereafter referred to as *Report 1963–64.*)

hoped that some 60,000 acres of rehabilitated palm groves would have been established in the Eastern Provinces.

What is required is a dramatic 'agro-surgical' operation, i.e. the cutting out of old, diseased and unproductive trees, and the replanting of the land to improved oil palm seedlings, as developed at NIFOR (the Nigerian Institute for Oil Palm Research) at Benin City in Mid-Western Nigeria. The felling of oil palms, even those of known low-yielding ability, means some loss of income to the villager; hence the Government subsidy scheme which reimburses tree owners at the rate of eighteen pounds per acre for the removal of the old trees and their replacement with high-yielding seedlings. Part of this grant is in the form of cash, the remainder in seedlings and fertiliser.

A farmer or a group of farmers must have a minimum of five acres of land to qualify for aid under the rehabilitation scheme. Ideally the land should be in one piece, although in practice two or more separate and near-contiguous plots may qualify for the programme. Removal of the old trees is undertaken, with the exception of twenty palms per acre which may be left standing and are intended to continue yielding the landowner some income at least over the following four years, until such time as the new palms have started to produce their first bunches of fruit. After four years, the remaining old trees are cut down and the entire area is planted to improved trees.

When fully mature, NIFOR palms will yield over six times as much palm oil, and three times as many kernels, as semi-wild palm groves. in terms of cash returns, while unimproved bush palms may yield about eight pounds an acre per year, a replanted palm grove may yield as much as forty-four pounds an acre, a more than five-fold increase. While the acreages involved thus far are small, and the growth rate modest, the exercise is catching the attention of many farmers, and a greatly increased rate of group participation may be expected over the remaining years of the programme. Response at the core of the oil palm belt, in Owerri and Annang Provinces, has also been encouraging.

Livestock. Domesticated animals have traditionally formed only a minor part of the rural economy of Eastern Nigeria. Small animals such as fowls, goats and sheep are kept on almost all compounds but their economic value is small. They are reared mainly for slaughter and consumption on ceremonial and festive occasions. The main utility of the small livestock is in the manure they produce for intensive compound-type farming in and around the village.

G F.E.N.

It is difficult to rear larger animals such as cattle, horses and other draft animals in Eastern Nigeria due to the prevalence of tsetse and the spread of trypanosomiasis. There are however considerable numbers of dwarf shorthorn cattle, the *muturu*, which have an immunity to the tsetse-transmitted disease, at least within the environs of their birthplace. These forest-originated livestock are most common today in the derived savanna lands of Enugu and Abakaliki Provinces; the southern part of Uyo Province also supports many small herds. Again their economic value as meat and milk producers is minimal; they have a social value in the prestige they give to their owners, also a socio-religious value as sacred beasts in certain traditional systems of belief.

Livestock developments in recent years have seen a strong emphasis, under American auspices, on poultry production for eggs and broilers. Hatching eggs and day-old chicks (Rhode Island Reds) are distributed from Abakaliki Government Farm and small-owner poultry farms are an increasingly common sight in many parts of the region. Increased egg production is making a major contribution to improving the diet of the people; however, the danger of decimation of flocks from poultry diseases has yet to be overcome. Pigs and rabbits are being promoted, while there are active programmes to improve the strains of goats and sheep.

Upgrading of cattle is also under way, with experiments on government farms and at the University of Nigeria involving controlled breeding and feeding of *muturu*, *Ndama* (a West African forest animal), and the white Fulani *zebu* cattle from Northern Nigeria. The major source of beef for Eastern Nigeria remains the north; under normal, peacetime conditions, large numbers of cattle move into the region, mainly 'on the hoof', arriving for slaughter in a poor if not emaciated condition after days and sometimes weeks on the road. In 1965 lines of slow-moving cattle with their Fulani or Hausa herdsmen were a common sight on authorised routes leading to the larger urban centres in the east. Most of the meat sold in the market-stalls of the region is low in quality and its supply is both limited and erratic.

In an effort to improve on this situation, the Eastern Nigeria Development Corporation is developing a livestock ranch on the Obudu Plateau, where at 5,000 feet excellent tsetse-free grazing on montane grasses is available for herds of Gudali and other breeds of *zebu* cattle.[1] Cross-breeding with Devon bulls has also been initiated, in order to establish a dairy as well as a beef industry. An area of approximately 40 square miles

[1] For a fuller description of the Obudu operation, see B. N. Floyd, *ENDC Ranch: Cattle Ranching in Eastern Nigeria* (Enugu, 1965).

or 25,600 acres is available for the ENDC operation, stocked in 1965 with over 3,500 head. With paddocking and improved pastures through seeding with exotic grasses, a maximum stocking rate of 4,000 cattle is visualised, with an annual offtake of 1,000 head. The cattle are shipped by truck a distance of 200 miles to Enugu for slaughter; a beef lot at Adada on the Nsukka Plateau is being used for fattening some of the cattle on forage crops, silage and grains before they are passed to the abattoir at nearby Oghe, a few miles from Enugu.

In association with the cattle ranch is the successful ENDC Obudu Ranch Hotel; the striking scenery and ameliorated climate of the plateau make of the area an excellent holiday resort, especially for expatriates who benefit both physically and psychologically from a stay in the cool uplands. Obudu stands to become well known in West Africa as a tourist centre with very real attractions.

11 Shortcomings of Traditional Agriculture and the Need for Innovations

INTRODUCTION

WE have already noted that agriculture is the mainstay of the Eastern Nigerian economy and that within the foreseeable future it will continue to be the region's major source of employment and income. We have also observed that many farming communities are finding it increasingly difficult to cope with the problems of supporting their peoples, let alone of producing saleable surpluses of agricultural products with which to finance much-needed and long overdue improvements in living standards. There is an obligation then 'to transform traditional agriculture, which is niggardly, into a highly productive sector of the economy'.[1] Those who farm as their forbears did cannot make the significant contributions to the economic growth of the Eastern Provinces which are necessary if a truly modern society is to emerge.

The crucial consideration concerns the degree to which traditional concepts of land use should be preserved or maintained in light of the mounting crisis in the rural areas. The question may be posed thus: are the customary socio-agricultural organisations of the rural population adequate to meet the challenge of the 'revolution of rising expectations' which has swept across the underdeveloped areas of the world? To a geographer, directly concerned with man and his association with the physical world around him, the answer is in the negative. The traditional systems of indigenous agriculture in this part of Africa, however appropriate to past social and physical circumstances, are quite unsuited to the radically different situation of the present and the future. Such systems of land use have persisted long after the environmental and social conditions which called them into being have disappeared, and they have proved incapable of accommodating themselves adequately to new factors in the human and physical geographical setting. Above all they are entirely unsatisfactory for a population which has doubled itself over the last two

[1] T. W. Schultz, *Transforming Traditional Agriculture* (New Haven, 1964), p. 4. (Hereafter referred to as *Agriculture.*)

decades and – despite the fearful death-toll of the Civil War – may repeat the process over the next twenty to twenty-five years.

In Eastern Nigeria, traditional land customs have been tested out for their survival value since the turn of the twentieth century, when the area came under the control of a colonial power; they have been unable to match up to the demands of an emergent and expanding rural society. Before the advent of the British, social structures and farming techniques had been evolved which insured that the Ibo, Ibibio, Efik, Ijaw, and others lived in harmony with their environment, at subsistence level. But irreversible processes were set into motion with the arrival of the European. The old order of society was unavoidably modified and weakened. The introduction of a commercial economy and Western thought patterns struck at the very heart of tribal *mores* and organisation, resulting in increasingly unharmonious relationships between man and land in Eastern Nigeria.

Thus the disintegration of traditional life commenced with the active British colonisation of Nigeria and the degree and tempo of acculturation have accelerated markedly in the successive decades of this century. Continued change would appear inevitable and it is beside the point to argue whether it is desirable or not. It must be accepted that essential improvements in farming techniques cannot take place without a radical alteration of the customary concepts of land use and tenure, necessitating the ultimate abandonment of many former practices. It follows that attempts to preserve the traditional ways of doing things must eventually succumb to the impact of modern economic forces.

Shifting cultivation, bush fallowing, indiscriminate use of fire, excessive fragmentation of holdings resulting from a claim to the land on the part of every adult male, these and other practices are relics of a tribal economy which are incompatible with effective development of agricultural resources in the Eastern Provinces. Moreover Eastern Nigerians aspiring to become successful farmers must somehow shake themselves free from the 'shackles of the ever-ramifying kinship system' with its traditional demands upon those who reap a bumper harvest. It is mere sentiment to extol traditional societies and to defend their ways of living when those ways now provide total earnings of only a few pounds a year. Preservation of land-use customs in the face of persistent poverty would appear to be an entirely unacceptable proposition.

Reduced to its simplest terms, the issue is essentially one of the preservation of natural resources versus the preservation of traditional systems: human and land development versus human and land degradation. A socio-rural 'revolution' is thus inescapable and is, in fact, long overdue.

While agreement may be reached on the necessity of basic agrarian reforms, it is another matter to know how to introduce these changes so as to cushion their impact and to create as little harmful disruption as possible to existing social patterns. There is a need to conserve goodwill as well as the soil; 'human husbandry' is just as significant as 'land husbandry'. Ultimately all the problems of land use centre around the farmer and his family. If his views are not adequately taken into account, if he is not won over to the programme of reforms which is planned on his behalf, then his ways of farming will never be improved. Land-use planners in the Ministries of Agriculture and Economic Development should never forget that 'it is the people who matter, not the soil, the crops or the livestock, except in their relation to the total prosperity and happiness of the people'.[1] In many ways, Eastern Nigerian farmers are already a long-suffering people; it would be fatal to regard their interest in the future as non-existent or indefinitely pliable.

Most fortunately, there has appeared in recent years a mounting interest in innovation and an eagerness to learn and to try out new ideas on the part of an increasing number of farmers in Eastern Nigeria. This appears to fly in the face of the established picture of marked conservatism and distrust of change on the part of rural populations. Indeed, it is axiomatic to state that farmers the world over are by nature conservative. 'According to urban folklore, agriculture is the Gibraltar of traditionalism.'[2] When one is living close to or on the poverty datum line, one is not unnaturally reluctant to experiment with new and untried methods of farming that may not succeed, and may bring the farmer and his family to the brink of starvation. At the same time, the idea that all cultivators are irrevocably against change is a fallacy. 'The notion that all farmers are handcuffed by tradition, making it impossible for them to modernize agriculture, belongs to the realm of myth.'[3] One does traditional cultivators an injustice by assuming that they will not respond to new ideas once their value has been clearly demonstrated.

In Eastern Nigeria, due to a vigorous rural education programme among other factors, many farmers have been quick to grasp the advantages which can be derived from new techniques and practices.[4] One need

[1] C. W. Lynn, *Agricultural Extension and Advisory Work*, with special reference to the Colonies, Colonial Office Pub. No. 241 (London, 1949), p. 104.

[2] Schultz, *Agriculture*, p. 162.

[3] Ibid.

[4] For a fuller treatment of cultural factors favouring agricultural change in Eastern Nigeria, see D. Smock, *Agricultural Development*, Pt. II, pp. 17–26.

not therefore anticipate a reactionary social response to the principles of agricultural reform, even if the specifics prove less easy to implement.

There are in fact some formidable human and physical constraints on the adoption of new systems of rural land use. These may be considered under the heading of:

(i) Land tenure;
(ii) environmental aspects;
(iii) economic problems.

Proposals as to how these handicaps can, and must, be overcome are included in each section.

(i) LAND TENURE[1]

Perhaps the greatest obstacle to a more rational utilisation of the land for agriculture stems from the traditional concepts of land ownership, which are in turn derived from the days of small self-supporting communities with ample land. In most Eastern Nigerian communities the land belonged, as it still does in the last analysis, to a group of kin, a family or a clan, the membership of which included not only the persons alive at any particular time, but persons dead and persons not yet born. Land was therefore more than tangible property; it expressed the social and spiritual identity of a group of kinsmen in contradistinction to other groups in other communities.

When mythology, magic, religion and sentimental attachment to ancestors and descendants is mixed with land tenure, the issue becomes a highly complicated one. For generations, these mystical ideas about the land have dictated when, where, how and by whom it should be used for agricultural purposes. Of special significance was the cardinal principle that no member of the community should be without land for compound, gardens and farms. Concepts of this kind die hard, and reforms in land use require in turn a psychological and spiritual reform within the land-owning groups.

Traditional communal land tenure may have been a wise adaptation for common security in former circumstances.

> But let there be no mistake; it is possible to appreciate, even with a certain admiration, the complexity of a culture so intricately adapted to its limitations, but the limitations are preposterous and the

---

[1] This important topic has been dealt with in a thorough fashion by L. T. Chubb, *Ibo Land Tenure* (Ibadan, 1961); T. O. Elias, *Nigerian Land Law and Custom* (London, 1962); and S. N. C. Obi, *Ibo Law of Property* (London, 1963).

culture is desperately poor. Better methods have still to defer to soil and climatic conditions but in many areas will demand a revolution in land tenure.[1]

In Eastern Nigeria changes have occurred over the years, to be sure, and there is today a strong measure of individual control of lands, particularly with regard to the vital issue of transfer of land. This may take place in three sets of circumstances.

(a) Seasonal. Land is made available for the use of another person for a single farming season, being either rented or donated as a traditional gift. It is interesting to note that measurement of land area is achieved by estimating the number of yams that can be planted on it. In Iboland the basic unit is 400 hills or heaps, and land is rented in subdivisions or multiples of this unit.[2]

(b) Temporary. Land is made available for an indefinite period but is ultimately recovered. The transaction is most commonly by means of pledging (or mortgaging), i.e. the granting of a piece of land to someone in return for a sum of money which, when returned, obliges the person to leave the land. Short-term leases on indefinite terms are also becoming more common in parts of Eastern Nigeria.

(c) Permanent or outright. Such a transaction may be unconditional or subject to reservations, and occurs either as a gift, as compensation for a wrong or, very occasionally, in exchange for money (sale). This third category of land exchange is the least common since it is largely true to say that individuals, or families, or whole villages, still do not have the right to sell or permanently alienate the land in their trust.

One of the main features of land tenure in Eastern Nigeria is the fractioning and fragmentation of holdings. Fractioning refers to the constant division and subdivision of plots over the years which has produced small, even diminutive parcels of land today. Fragmentation implies the scattering of plots, to the point where a farmer may control fifteen to twenty pieces of farmland, separated by distances of up to three miles. Some farmers may have half this number of plots under simultaneous cultivation.

Fractioning and fragmentation are a result of the system of inheritance whereby a farmer with holdings covering different soil types and suited for

---

[1] G. Hunter, *The New Societies of Tropical Africa* (London, 1962), p. 101. (Hereafter referred to as *New Societies*.)

[2] G. I. Jones, 'Ibo Land Tenure', *Africa*, xix (1949), p. 318.

different crops will, to be fair to his sons, leave them each a share of all his plots.

Such fragmentation constitutes a formidable barrier to agricultural development, for it makes efficiency of production very difficult. For one thing, it takes the farmer considerable time to move from one plot to another. As a result he cannot care for his crops as he should Moreover, it is very difficult to transport seedlings, fertilizers, and other items to his various fragments and to collect and transport his products to market. In fact, it is virtually impossible to move lorries to a man's various fragments in order to deliver supplies or to collect products because to lay out roads so that each plot is on a road would mean using about half the available land for roads. Large-scale improvements by a farmer to his land, such as terracing, or setting up drainage or irrigation systems are almost unthinkable when dealing with so many little, scattered plots. Also, the introduction of mechanization is practically pointless under such conditions.[1]

Another barrier to agricultural development posed by the tenuriae system is the great difficulty of obtaining new land. Even if a farmer has the funds to acquire and develop additional land, it is very difficult for him to do so both because of the security of land and the manner in which land is held. In most rural areas of the Eastern Provinces, as indicated, the notion of selling land outright is rare. When it happens, it is invariably between people of the same village for building purposes rather than for farming. Land may be obtained through pledging, but it is of little value in terms of serious agricultural development since it can be reclaimed by the holder at any time, on repayment of the pledge; also in most instances only annual crops may be raised. Once trees are planted, the customary owner fears that the temporary occupants may try to claim possession of the land. Since tree crops appear to offer the best investment for commercial agriculture, this limitation is a grievous one. Indefinitely-termed leases on surplus land are no better than pledges due to a similar uncertainty over the length of time the land may be cultivated.

Another feature which inhibits agricultural reform is the suspicious attitude adopted by communities with land to spare towards 'strangers' or outsiders who wish to acquire farmland. While they may be willing to accept the strangers as rent-paying tenants, there is considerable reluctance to let outsiders obtain control over land through gift, pledge or sale.

Under customary land tenure, there is no legal title deed or land registration. Orally-transmitted tradition usually backs up a farmer's

---

[1] Smock, *Agricultural Development*, p. 28.

claim to his land but land disputes occur not infrequently, putting large tracts of land out of cultivation as a result of the controversy. Due to the inevitable social upheaval and occasional violence, leading to loss of life, land disputes often end up in the courts, where sizeable sums of money are spent on litigation which might far better have gone on development of the disputed land.

Absence of registration certificates or title deeds also means that farmers find it difficult to use the land as collateral for farm improvement loans from banks and other lending agencies. This acts as a further deterrent to land development on the part of enterprising farmers.

In summary, many aspects of traditional tenure act as serious constraints to the modernisation of agriculture in Eastern Nigeria. There is urgent need for tenurial reforms so that the breakthrough to rational farming units and land-use systems can occur.

The greatest need is for the merger or consolidation of the bits and pieces of fragmented land into viable blocks. Although fragmentation may have served as an equable form of social security in the past, its persistence to the present day thwarts any serious attempt at establishing economic holdings. A region-wide programme of land consolidation has yet to be initiated but should receive high priority. Undoubtedly there would be strong reactions from most communities, and many objections raised concerning the difficulties of redistributing the land fairly, ancestral disapproval, and so forth. But in this connection, the lessons learned in other African countries where this essential exercise has been undertaken, e.g. in Kikuyuland in Kenya, are directly relevant to Eastern Nigeria.

The technological and economic advantages of centralised, single-unit farms for establishing intensive systems of commercial farming are undeniable. Whether the blocks are small and individually owned and operated, or large and farmed on a communal or co-operative basis, the value of consolidation soon shows in the increased facility with which modern techniques of soil and crop management, also efficient marketing of products, can be implemented.

Securing titles to consolidated farms is also essential so that long-range beneficial procedures and tree crops can be adopted. Where outright ownership of the land is not possible, the acquisition of land through long-term leases would prove a satisfactory solution to the problem. Official title deeds or registration cards should be issued and as valid security for agricultural credit with which essential improvements can be funded.

Lastly it must be appreciated that henceforth not every adult male can expect to own farming land in his village of birth. Alternative means of livelihood and forms of security must be developed other than those which stem from cultivating minute and scattered patches of soil for a few months of the year. But this notion will prove the most difficult of all to accept until such time as attractive alternatives become a reality, and this depends on overall economic planning involving country and town, rural and industrial occupations.

(ii) ENVIRONMENTAL ASPECTS

The soils of Eastern Nigeria, whether ferralitic or ferruginous, hydromorphic or youthful in origin, have one thing in common: an inability to withstand annual cropping under rudimentary sedentary techniques of cultivation. Even the intensive compound management system, with regular additions of organic compost or manure, achieved only through large labour inputs, barely manages to maintain the productivity of the soil.

Modern soil conservation measures should be adopted on a region-wide scale as quickly as possible. They involve the replenishment of soil nutrients through regular applications of chemical fertilisers and organic matter; the adoption of scientific crop rotations; the establishment of groves of tree crops; and the mechanical protection of soil from run-off and erosion through the overlay of a planned network of contour banks and grassed drainage channels. Rearing of small livestock under scientific management conditions should also be undertaken. These ambitious proposals are, of course, difficult to implement immediately and in their entirety for a number of reasons. Land consolidation and security of tenure should precede infrastructural improvements to the farm in terms of construction of contour terraces, fencing, new buildings and improved storage facilities. The use of fertiliser assumes capital available for its purchase; adoption of new crops in rotation requires knowledge of their culture; acceptance of these innovations requires further mass-educational programmes in the rural areas. But in terms of the above changes a good start has already been made in persuading the people 'to want what they need'. The Agricultural Information Division of the Ministry of Agriculture (through publications, film shows, radio programmes and lectures), the Community Development and Co-operative Divisions, the agricultural extension services (especially the Soil Conservation Section), demonstration farms, leadership-training programmes, workshops and research at Umudike and the University of Nigeria, Nsukka: these and other agencies have provided the catalyst whereby a proper climate of opinion towards needed changes has been created.

Fertilising the Soil. Enrichment of soils which have been depleted due to over-cropping may be accomplished by applications of organic plant and animal wastes, or by the use of chemical or 'artificial' fertilisers.

The collection of kitchen wastes, compound sweepings, chicken and goat droppings, leaves and grasses, has provided the traditional way of making manure in Eastern Nigerian villages. This procedure and the knowledge of the benefits which it renders may be built upon to encourage farmers to produce far greater quantities through battery composting (the Indore method) and mixed farming practices.

Chemical fertilisers represent an unfamiliar way of maintaining soil productivity; due chiefly to expense, but also to ignorance concerning methods of application and suspicion regarding their effect on the taste and storage qualities of yam and other tubers, mineral fertilisers have not as yet been adopted on any wide scale. But such an eventuality is essential for the future of farming in Eastern Nigeria. Research in Nigeria and elsewhere has shown beyond doubt the agronomic advantages to be gained from the use of chemical additives, especially sulphate of ammonia, superphosphate, potassic fertilisers, applied either singly or in mixes (NPK).

Among other bodies, the Norwegian Church Agricultural Project (NORCAP) at Ikwo, near Abakaliki, has carried out fertiliser trials on staple root and cereal crops. Results for selected crops are shown in Tables 11.1 and 11.2.

TABLE 11.1

Norcap fertiliser trials 1963: Cassava. NPK 13:13:18 mix[1]

Varieties	Mean Yields (tons per acre)		% Difference
	With fertiliser	Without fertiliser	
513101	4·8	3·6	33
Jiakpu	4·3	3·3	30
44086	3·9	2·5	56
37065	3·6	2·4	50
Olube	3·0	1·6	88
43083	2·9	2·1	38
Panya	2·7	2·4	13
Onu Ocha	2·5	1·6	56
AVERAGE	3·5	2·4	46

[1] O. Aurlien, 'Research Results from Fertilizer Trials', paper presented at Conference on Fertilisers, University of Nigeria, Jan. 1964, p. 1.

TABLE 11.2

Norcap fertiliser trials 1963: Yam. Topdressing of NPK 13:13:18 mix[1]

I – *No fertiliser before planting*	Yield Per Acre *(Tons)*
No topdressing	2·94
Topdressing (one application)	4·56
Topdressing (three applications)	4·90

II – *Fertiliser mixed in the soil before planting* (400 lbs/acre of NPK mixture)	
No topdressing	6·54
Topdressing (one application)	6·70

Experiments by the Ministry of Agriculture show similar impressive increases in crop yields. The deep porous sandy soils of the southern and central areas are low in nitrogen and potash but tend to have sufficient phosphorus. Heavy treatments of NPK increase yam yields appreciably. Soils in the north-eastern part of the region respond to phosphatic fertiliser.[2]

Use of chemical fertilisers for tree crops has been the practise on plantations for many years, and the benefits in terms of accelerated growth of seedlings, increased yields of oil palm and cocoa, and resistance to disease are clear-cut.[3] Their use by smallholder tree-crop farmers will also be essential in the future.

The economics of chemical fertilisers raise perhaps greater problems than their application, although a good deal of experimental research is still required to ascertain the precise needs of different soils and crops at field levels.[4] The purchase costs of fertilisers must be more than offset by the increased yields and value added to the crops produced. Government-sponsored subsidy schemes are helping to popularise fertilisers and should be greatly expanded. Such subsidies are a temporary expedient to build the demand for fertilisers to a point where prices will fall, allowing the subsidies to be reduced and eventually abandoned.

Another area where the chemical industry may be expected to assist the agrarian 'revolution' in Eastern Nigeria is in the control of pests and diseases which at the moment take a heavy toll of harvests. Insecticides, pesticides and fungicides will need to be used on a much larger scale than

[1] Ibid., p. 2.
[2] Eastern Nigeria, Ministry of Agriculture, *Report 1963–64*, p. 59.
[3] J. M. A. Sly, 'Research Results from Oil Palm Fertilizer Trials WAIFOR', paper presented at Conference on Fertilisers, University of Nigeria, Jan. 1964.
[4] G. L. Johnson, 'Making Practical (Economic) Fertilization Recommendations', paper presented at Conference on Fertilisers, University of Nigeria, Jan. 1964.

at present, where their use is restricted to the larger plantations of external exchange crops.

Crop Rotations. The growing of food and cash crops in rotation, with the inclusion of legumes and a short-term grass fallow or ley, is a further requisite for keeping the soil in good heart. Elimination of lengthy periods of natural regeneration under bush requires, in addition to fertilisers, a sequence of crops which will make full use of the soil's capacities, contributing at the same time to its maintenance and productivity.

A two- to three-year fallow period with the soil occupied by a useful cover crop is considered the best substitute for a longer rest under natural fallow. A plant particularly recommended is *stylosanthes gracilis*, a versatile legume which may be turned under as a green manure crop at the end of the fallow phase, also utilised during growth as a livestock feed. Other fallow crops include molasses and elephant grass, Southern Gamba, *pueraria* and *centrosema*.[1]

A six-year basic food-crop rotation is being recommended for farmers in the sandland areas of the region, although local variations in topography, drainage and soils may require modifications to the sequence:

First year – yam, inter-cropped with early maize (to be harvested before the yam matures) and inter-planted with cassava (to be harvested after the main crop has been removed).

Second year – cassava, followed by pigeon peas or 'stylo' fallow. Pigeon peas which are also counted on as a substitute for bush fallowing, have the additional advantage of being edible by humans and a source of high protein (they are sometimes referred to as 'poor man's meat').

Third to sixth years – Pigeon peas or stylo fallow.

Mechanical Soil Conservation Measures. The need to protect farmlands from the destructive effects of the elements, particularly precipitation, is widely recognised. A grid pattern of banks or terrace walls (bunds) aligned along the contour is the most commonly accepted measure for protecting fields and blocks of crop land from erosion. The contour banks serve to arrest the downhill surface flow of water and, by means of shallow ditches on their upper slopes, lead it safely off to grassed drainage channels, or allow it to soak into the soil *in situ* (Fig. 11.1).

The indigenous development of terraces by the Maku farmers in response

---

[1] Eastern Nigeria, Ministry of Agriculture, *Report 1963–64*, p. 71.

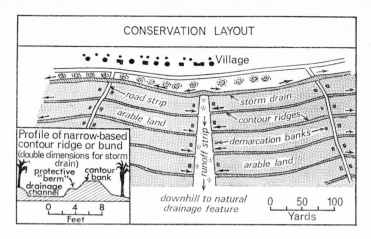

CONSERVATION LAYOUT

Village

road strip

storm drain

arable land

contour ridges

demarcation banks

runoff strip

arable land

Profile of narrow-based
contour ridge or bund
(double dimensions for storm
drain)

protective
"berm"

contour
bank

drainage
channel

0   4   8
Feet

downhill to natural
drainage feature

0   50   100

Yards

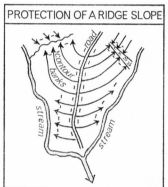

PROTECTION OF A RIDGE SLOPE

road

contour
banks

stream

stream

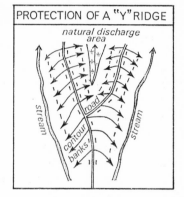

PROTECTION OF A "Y" RIDGE

natural discharge
area

stream

road

contour
banks

stream

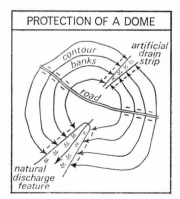

PROTECTION OF A DOME

contour
banks

artificial
drain
strip

road

natural
discharge
feature

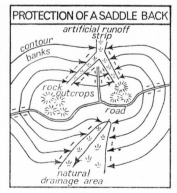

PROTECTION OF A SADDLE BACK

artificial runoff
strip

contour
banks

rock
outcrops

road

natural
drainage area

FIG. 11.1   *Mechanical soil conservation measures*

to their hilly environment requires widespread emulation across the Eastern Provinces, even on gently sloping lands which may appear not to be subject to sheet erosion from running water. While crude in layout, the conservation principles incorporated within the Maku system are sound.

The overall layout of the protective grid requires the services of professionally-trained soil conservationists, and accurate alignment of the bunds along the contour calls for trained levellers. Teams of conservation surveyors should be available in all provinces and divisions of the region for the demarcation of these features, particularly when schemes for the consolidation of lands start to produce tangible evidence of merging plots to be placed under permanent tillage.

(iii) ECONOMIC PROBLEMS

Implicit in all the above plans for transforming traditional agriculture is the assumption – a very big one – that sufficient capital will be available for mounting the required programmes of reform.

Large-scale expenditure is required for developing the regional infrastructure of communications: roads; bridges; port facilities; the railway; by which agricultural produce may flow more freely from the producing areas to domestic and overseas centres of consumption, and a reverse flow of goods and services may enter the rural areas for promoting development. Agricultural extension services, involving both large- and small-scale demonstration units, mechanised equipment and trained staff, require expanding, as does agricultural research on crops and livestock. New industrial plants for processing agricultural products are required. The list can be extended without difficulty. Individual farmers require more modest sums for investing in improved seeds and fertiliser, sprays and fumigants, upgraded livestock, better tools such as simple hand-operated machines for sowing seeds with fertiliser, small 'barn' machinery for processing crops, new buildings, and the hiring of labourers when necessary for erection of conservation works.

The capital requirements of agriculture are but a part, although the single most important part, of the overall investment needs of Eastern Nigeria (Chapter 18). Dependence upon external agencies and investors is unavoidable to meet these financial requirements since internal sources for funding are limited and domestic savings (due to low *per caput* incomes) negligible. The World Bank, the governments of the Western nations, the United Nations, International Foundations, private companies and investors, and other bodies, are the principal sources of development financing. The biggest thrust of the United States' Agency for International Development (USAID) has been in the realm of agriculture.

Within Eastern Nigeria, beside ministries and government departments, there are a number of institutions which aim – either directly or indirectly – at providing development resources for the rural population. These include Community Development, Co-operatives, the Eastern Nigeria Marketing Board, the Eastern Nigeria Development Corporation (ENDC) and the Fund for Agricultural and Industrial Development (FAID). There are also a number of church-related projects similar to that of NORCAP. Community Development programmes and the work of the co-operatives both help to mobilise unused or under-utilised resources in the villages for the welfare of the community as a whole; each helps the villages to help themselves. The Community Plantations Project is discussed in Chapter 12. The Marketing Board through its policies directly affects hundreds of thousands of farmers; it is also a means by which development resources are mobilised. The ENDC is an instrument for direct governmental initiative and participation in specific agricultural and industrial projects for the development of the economy. The ENDC plantations are described in Chapter 12. FAID provides loans to individuals for the development of their farms.

It will be necessary for all these agencies to expand their operations significantly in the years to come. Co-operatives in particular should extend their functions, assisting for example with the establishment of new cash crops such as tobacco, pineapple, market-garden crops, as well as the marketing of traditional products.

> Co-operative Societies are the main institutions providing credit for village agriculture, for petty trade, and for cottage and handicraft industries, that is, for the occupations on which well over three-quarters of the population of the Eastern Region depend for their livelihood. The growth of the co-operative movement has been slowed by the inevitable difficulty of creating necessary working capital from the savings of members who are living essentially on a subsistence level. . . . Despite its small size we believe that co-operation is rooted soundly; it should be used as the channel for the injection of Government funds and for the dissemination of new methods and skills necessary to expand the production of small-scale agricultural and industrial enterprise.[1]

---

[1] Ford Foundation, *Prospects and Policies for Development of the Eastern Region of Nigeria* (Enugu, 1960), p. 80.

The Eastern Nigeria Marketing Board pays guaranteed prices for oil palm products, cocoa and other crops; it provides a means of price stabilisation which benefits the small producer, even if the guaranteed price is consistently maintained at a level somewhat below average market price. In fact the Board levies a tax on the producers of the crops which it handles, and not enough of the profits (which have been very large in the past) have found their way back into investments and funds which might be of direct benefit to the farmers.

The FAID Scheme is in principle sound but again not enough of its resources have reached to the 'grass-roots' level of the impoverished but energetic peasant farmer who wishes to embrace new practices of cultivation. Too many beneficiaries have been local chiefs or well-to-do influential persons in a community: the people whom it is less necessary to help but who are able to offer the required collateral.

The formidable task of raising and distributing funds within the agricultural sector for attaining the breakthrough to scientific farming cannot be gainsaid. Upon the efforts of the Government and its planners hangs the destiny of the rural population in Eastern Nigeria.

12 Modern Developments in Agriculture

INTRODUCTION

WE may now review some of the agricultural projects which have been launched or greatly expanded by the Government since independence, and which are intended to spearhead the agrarian 'revolution' in Eastern Nigeria.

An earnest of the Government's determination to effect dramatic changes in rural land use is provided by the estimate of capital expenditure on agriculture in the Eastern Nigeria Development Plan. Some £30,361,000, or 40 per cent of the estimated total expenditure of £75,192,000, was earmarked for primary production (agriculture, forestry, fisheries), the lion's share of which is for agricultural projects. No other sector of the economy is allocated so high a percentage. The figure was revised in 1964, to £27,896,000 or 37 per cent of the total expenditure, but agriculture still occupies pride of place in the plan.[1] Altogether some thirty different projects have been launched in a massive assault on poverty and backwardness in the rural areas of the Eastern Provinces. There has been a realistic decision to try to penetrate the traditional fabric of village life on several fronts.

Generally speaking, the schemes fall into two broad categories:

(1) Fundamental improvements to prevailing systems of land use based largely on traditional tribal tenure and indigenous settlement patterns. These are being introduced gradually by persuasion, by rural education, propaganda and demonstration and through exercising economic incentives such as the manipulation of market prices and the granting of subsidies. The Tree Crop Subsidy Scheme, including the Oil Palm Rehabilitation Scheme described in Chapter 10, also the Community Plantations, to be discussed in the present chapter, fall into this category of agricultural betterment schemes.

---

[1] Nigeria, Federal Ministry of Economic Development, *National Development Plan. Progress Report 1964* (Lagos, 1964), Table 4·3, p. 63.

(2) The legislative introduction of more novel, often alien, systems of land use, involving radical changes in tenure and social organisation; these innovations are occurring with some speed and are based on the assertion of political powers which both reorganise agricultural productivity and administer farming activities, with no small measure of authority. This is the so-called 'command' approach to the transformation of traditional agriculture.[1] Illustrative of these schemes are the Plantation Projects of the ENDC, and the Farm Settlements, both of which are examined in this chapter.

PLANTATION AGRICULTURE

The term 'Plantation' implies an organisational approach to agriculture which is quite different from small-scale, subsistence farming by individuals. In the Western sense, a plantation is a highly capitalised system of land use for the production, usually for the world market, of one or more tropical cash crops. Land, measured in thousands of acres, labour, processing and marketing are efficiently organised along industrial lines.

Plantations have thus far played a very minor role in the agricultural history of Eastern Nigeria. The colonial authorities discouraged Western-owned and operated plantations on the grounds that they were socially disruptive and exploited the indigenous labour force.[2] In the 1920s and 1930s, William Lever (later Lord Leverhulme) was continually refused permission to establish oil palm plantations in Southern Nigeria and turned instead to the Belgian Congo where flourishing estates were initiated; the high-grade products from these plantations are now challenging the largely peasant-derived oil palm products of Eastern Nigeria. Efficiently-run oil palm plantations were also established at this period in the Dutch East Indies. From the global viewpoint, the industrialised nations required a large and reliably regular supply of good-quality palm oil and kernels; they were not concerned with the area of origin or the methods of production. In the Southern Provinces the British administration attempted to promote smallholder production through the distribution of improved seedlings, establishing marketing boards, and offering enhanced prices for higher-quality products.

This policy continued until 1951. With political advancement, and the creation of the first Cabinet of indigenous administrators in 1952, the

---

[1] Schultz, *Agriculture*, p. 102.

[2] For a fuller treatment of the historical development of plantations, see R. K. Udo, 'Sixty Years of Plantation Agriculture in Southern Nigeria, 1902–1962', *Economic Geography* xli (1965), pp. 356–68.

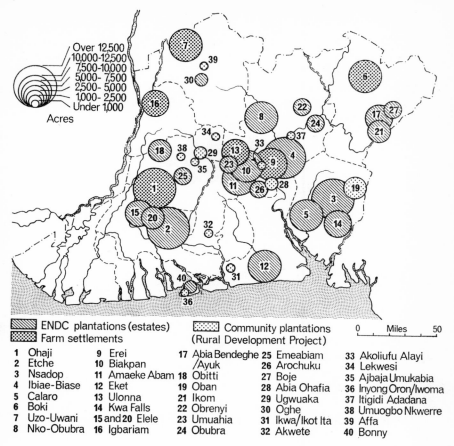

Key:

- ENDC plantations (estates)
- Farm settlements
- Community plantations (Rural Development Project)

Scale: 0 Miles 50

Legend (Acres):
- Over 12,500
- 10,000–12,500
- 7,500–10,000
- 5,000–7,500
- 2,500–5,000
- 1,000–2,500
- Under 1,000

1	Ohaji	9	Erei	17	Abia Bendeghe /Ayuk	25	Emeabiam	33	Akoliufu Alayi
2	Etche	10	Biakpan	18	Obitti	26	Arochuku	34	Lekwesi
3	Nsadop	11	Amaeke Abam	19	Oban	27	Boje	35	Ajbaja Umukabia
4	Ibiae-Biase	12	Eket	21	Ikom	28	Abia Ohafia	36	Inyong Oron/Iwoma
5	Calaro	13	Ulonna	22	Obrenyi	29	Ugwuaka	37	Itigidi Adadana
6	Boki	14	Kwa Falls	23	Umuahia	30	Oghe	38	Umuogbo Nkwerre
7	Uzo-Uwani	15 and 20	Elele	24	Obubra	31	Ikwa/Ikot Ita	39	Affa
8	Nko-Obubra	16	Igbariam			32	Akwete	40	Bonny

FIG. 12.1 *Organisation of plantations*

closed policy of the British was reversed, ushering in what has been termed the 'plantation decade'. Unilever was now permitted to develop an oil palm estate near Calabar, while the largest expatriate concern, the Dunlop Rubber Estate, with over 21,000 acres at its disposal, began planting operations in 1956[1]. Many smaller plantations for oil palm, rubber, cocoa, cashew and coconuts were initiated under the direction of the quasi-governmental Eastern Region Produce Development Board (forerunner of the Eastern Nigeria Development Corporation).

[1] Due to a certain odium attached to the historic term 'plantation', they were given the euphemistic official title of 'estates'.

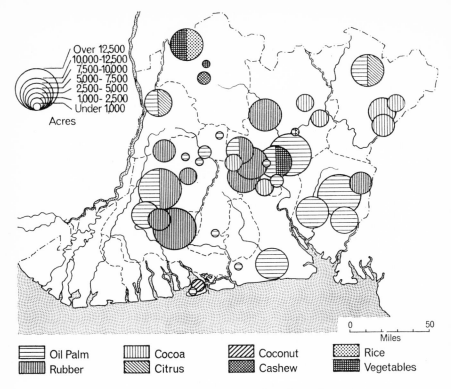

Over 12,500
10,000-12,500
7,500-10,000
5,000- 7,500
2,500- 5,000
1,000- 2,500
Under 1,000

Acres

0   50
Miles

Oil Palm   Cocoa   Coconut   Rice
Rubber   Citrus   Cashew   Vegetables

FIG. 12.2 *Principal crops produced on plantations*

The present distribution of plantations or estates and an indication of their organisational pattern, size and speciality crops is given in Figs. 12.1 and 12.2. Many of the ENDC plantations are less than half planted up, and the entire acreage of plantations in the region amounts to 1·3 per cent only of the total area of Eastern Nigeria (Table 12.1).

TABLE 12.1

Areas under plantation-type agricultural schemes in Eastern Nigeria, 1965 (gross acreages)

Eastern Nigeria Development Corporation	148,930
Farm Settlements	61,770
Foreign Companies	31,730
Community Development Projects	11,750
	254,180 acres

The plantation's contribution to the agricultural economy is similarly minimal due to the late start: some 5 per cent of the rubber and $3\frac{1}{2}$ per cent of the oil palm exports.

Active encouragement is given to foreign investors to establish plantations; they are classed as pioneer industries, thus qualifying for income-tax concessions including tax 'holidays'. A recent concept is the 'nucleus' plantation, a managerial system whereby commercial crops are produced in a core plantation area, managed as usual by hired labour and equipped with processing machinery; surrounding traditional communities are then encouraged to bring their commercial crops for processing to the nucleus plantation, and to emulate the superior production techniques of the parental agricultural unit. Dunlop has offered to assist a smallholder scheme around their large plantation at Uyanga, thirty-three miles north east of Calabar, while the Commonwealth Development Corporation is sponsoring the establishment of a nucleus plantation in Enyong Division. In 1966 the CDC announced grants totalling £500,000 to the Eastern Nigeria Nucleus Rubber Estate Company and a further £400,000 to the Eastern Nigeria Smallholder Board of Management for an adjacent smallholder scheme. The latter will involve some 4,000 acres of rubber in eight-acre plots for purchase by individual farmers in instalments over fifteen years; there will be about 2,500 acres of food crops and additional areas for villages, roads and amenities. The nucleus rubber estate will involve some 4,000 acres all told, and include a factory to process rubber produced both by the estate and by the smallholders.[1]

We may now look more closely at three illustrations of plantation-type developments in the region:

(i) community plantations;
(ii) the ENDC plantations or estates;
(iii) the farm settlements.

(i) COMMUNITY PLANTATIONS

These plantations are at the heart of the Eastern Nigeria Rural Development Programme which seeks to translate into reality, at the grass-roots level of the village, the objectives of the Six-Year Development Plan. It aims at the acceptance of new work patterns at the village level; at devising ways and means whereby those who actually farm can enjoy security of tenure; at consolidating fragmented holdings into more viable units; at

---

[1] Barclays Bank DCO, *Overseas Review* (May 1966), p. 66.

changing and modernising village life *in toto*. Maximum use of local resources is sought in these endeavours, capitalising on the enthusiasm for co-operative effort which in the past was the basis for such community projects as the building of roads, bridges, schools and maternity hospitals.

Self-help is the keynote of the Rural Development Programme. It is felt that over-lavishness of outside assistance can kill the community development spirit: on the other hand, government lacks the resources completely to finance economic expansion throughout the region. There is a definite financial limit to the extent to which, for example, the ENDC plantations and farm settlements can be supported. Community development projects therefore plan to utilise abundant labour as a substitute for scarce capital.

In 1965 work began on the establishment of twelve community plantations. The location of these projects is shown in Figs 12.1 and 12.2, while fuller details are given in Table 12.2. In each area, the people are expected to release a sizeable portion of the land for development as a plantation. The block is officially surveyed and the farmers who will work the land are formed into a co-operative. Through a lease arrangement, the co-operative is given use of the land for a period of time ranging from sixty to ninety-nine years. Rent is paid either to the landowners or the community.

Each farmer is given control over a piece of land to develop, the size of which will vary according to the overall acreage and the number of participants. The allocations have been between six and ten acres in those plantations already initiated. In addition to clearing the bush, planting the cash crops of oil palm and rubber, weeding, conservation work, etc., the farmer plants food crops for himself and his family. The oil palm and rubber seedlings, also artificial fertilisers, are supplied free under the Tree Crop Subsidy Programme.

TABLE 12.2

Community plantations sponsored by rural development project, 1965

Location	Division and Province	Size (Gross Acres)	Approx. No. of Participants	Period of Lease (Years)[1]	Crops	Pop. Density Zone
1. Oban	Calabar (Calabar)	approx. 5760	50		Rubber	Low
2. Abia Ohafia	Bende (Umuahia)	1500	170	99	Oil Palm	Low–Medium
3. Ugwuaka	Okigwi (Owerri)	1500	30		Oil Palm	Medium

4. Ikwa and Ikot Ita	Opobo (Uyo)	700	145		Oil Palm	Medium
5. Akwete	Aba (Umuahia)	530	95	99	Oil Palm	Medium
6. Inyong Oron and Iwoma	Opobo (Uyo)	420	100	99	Oil Palm	Low
7. Ajbaja Umukabia	Okigwi (Owerri)	400	100		Oil Palm	Medium–High
8. Akoliufu Alayi	Bende (Umuahia)	400	100	99	Oil Palm	Low–Medium
9. Lekwesi	Okigwi (Owerri)	400	70		Oil Palm	Medium
10. Itigidi and Adadana	Afikpo (Abakaliki)	185	35		Oil Palm Rice	Low
11. Umuogbo Nkwerre	Orlu (Owerri)	90	25	60	Oil Palm	High–very High
12. Affa	Udi (Enugu)	25	50		Vegetables (tomatoes, onions, peppers)	Medium

11,750 acres   970 participants

[1] Where not recorded, the period of lease was still under negotiation in 1965.

Proceeds from the cash crops upon maturity will go directly to the farmer. The entire development is supervised by Rural Development Officers and specialist extension officials.

The advantages of this approach to agrarian reform are self-evident. The community plantations act first of all as tangible demonstration areas for other communities who may wish to follow their example. As Table 12.2 indicates, the majority of the plantations are established in the medium or medium-high population density zones, and can thus be visited by farmers from many surrounding villages who are interested in hearing at first hand and seeing with their own eyes how other communities are coping with common problems of rural development. Next, it is relatively inexpensive to set up a community plantation; the only direct financial assistance from the Government is in the tree crop subsidies and the salaries of officials to supervise the programme.

Moreover, because the participants remain in their home villages, it has been possible to avoid the costs of resettlement and the establishment of new villages which are often involved in plantation schemes. Also, the farmers continue to feed themselves through growing their own food crops, and this avoids the cost of maintenance. The farmers are able to grow their own food because they are only expected to plant up between one and two acres of the plantation crop each year, and once his portion of the plantation has been completely planted, a farmer will be able to maintain his four to ten acres without cutting down substantially on his food crop production.[1]

The consolidation of holdings makes supervision simpler. All participants are planting the same crops on contiguous pieces of land, in a similar manner to the farmers of Maku. In one of the project villages, as many as 170 farmers are involved, linked together in a Farmer's Co-operative. If these men were working on scattered holdings and growing a variety of crops, proper guidance would be much more difficult. Another community plantation is located on land which is owned by 400 different people within the village. The lease arrangements made in connection with this 'operational area' have enabled the farmers to consolidate their scattered fragments of crop land into viable units which can be worked far more efficiently and profitably.

Above all, this approach to rural development helps farmers to overcome one of the greatest difficulties hindering rational land use, namely the unavailability of land for tree crops. Villages unwilling to allow communal land to be used by individuals for planting trees have allowed the land to be used for commercial plantations, since title to the land is registered in the name of the community and the income from rentals is used for the benefit of the entire village for financing other projects such as water supplies, roads and clinics.

Indications thus far are that the community approach to agrarian reform, while not the only useful device, holds much promise for the Eastern Provinces. Community development combines an appeal to both traditional and modernising emotions: it harnesses the old tribal pattern of co-operative effort to the new desire for improvements and efficiency. There are nevertheless certain drawbacks to be noted.

> ... its reference to old traditions, and the reliance on enthusiasm, despite their present success, may well become weaknesses in future. There is little doubt that, in the long run, traditional communal effort will be more difficult to arouse; it will begin to die with the conditions of tribal life and subsistence agriculture which gave it birth. ... Enthusiasm, too, is a wasting asset. It was relied upon in the early days of the Russian Revolution, and it inspired the Jugoslav railway builders, with shovels in their hands and school books in

---

[1] Smock, *Agricultural Development*, p. 41.

their pockets. In Europe, it had a life of five or ten years; we have yet to prove that it will last longer in Africa.[1]

Yet the spirit of community effort among the Ibo and other ethnic groups of Eastern Nigeria has persisted well into the present day from the time it was first utilised for economic development, in the 1930s. It received new impetus in the surge of energy and enthusiasm following independence in 1960. Perhaps this remarkable spirit of co-operation will last long enough for the breakthrough into modern and more prosperous rural societies. Meanwhile the government must avoid the temptation to misuse and thus destroy this enviable sense of community.

(ii) ENDC PLANTATIONS OR ESTATES[2]

The ENDC has since 1960 been mounting a vigorous programme of plantation development along orthodox commercial lines. A total of 148,930 acres in twenty-two locations has been acquired through long-term leases for plantation projects. Of this area, some 67,000 or 45 per cent had been planted up by the end of 1965. A total of 126,820 acres or 85 per cent were to have been planted by the end of the Development Plan period (1962–8). The location of the ENDC plantations is shown in Figs. 12.1 and 12.2, while details of the estates are recorded in Table 12.3.

TABLE 12.3
ENDC plantations, 1965

Location	Division and Province	Size (Gross acreage)	Approx. acreage planted up by end 1965	Approx. No. of Participants		Pop. Density Zone
				Supervisory Staff	Labourers	
I. *Oil Palm Estates*						
1. Nsadop	Calabar (Calabar)	14,000	1,053	22	253	Low
2. Ibiae-Biase	Biase (Uyo)	13,767	2,998	38	728	Low
3. Calaro	Calabar (Calabar)	11,750	10,275	131	1,466	Low
4. Eket	Eket (Uyo)	9,000	2,655	42	460	Medium

[1] Hunter, *New Societies*, p. 111.
[2] This section draws on materials from a manuscript by C. I. N. Ogbonnaya, 'A Geographical Appraisal of the Eastern Nigeria Development Corporation's Plantations' (Nsukka, 1966).

	Location	Division and Province	Size (Gross acreage)	Approx. acreage planted up by end 1965	Approx. No. Supervisory Staff	Labourers	Pop. Density Zone
5.	Kwa Falls	Calabar (Calabar)	7,344	4,078	84	413	Low
6.	Elele	Ahoada (Port Harcourt)	7,000	6,386	67	1,016	Low
	TOTALS		62,861	27,445	384	4,336	

II. *Rubber Estates*

	Location	Division and Province	Size	Planted	Staff	Labourers	Zone
7.	Etche	Ahoada (Port Harcourt)	14,000	3,140	29	636	Low
8.	Amaeke Abam	Bende (Umuahia)	10,000	2,416	30	641	Low–Medium
9.	Biakpan	Biase (Uyo)	10,000	2,300	25	712	Low
10.	Nko-Obubra	Obubra (Abakaliki)	10,000	2,432	27	648	Low
11.	Obitti	Owerri (Owerri)	6,000	1,700	16	400	High
12.	Elele	Ahoada	5,364	3,144	37	687	Low
13.	Emeabiam	Owerri (Owerri)	3,153	2,800	33	520	High
	TOTALS		58,517	17,932	197	4,244	

ENDC plantations, 1965

	Location	Division and Province	Size (Gross acreage)	Approx. acreage planted up by end 1965	Approx. No. of Participants Supervisory Staff	Labourers	Pop. Density Zone
III. *Cocoa Estates*							
14.	Abia-Bendeghe Ayuk	Ikom (Ogoja)	6,660	3,514	41	466	Low
15.	Ikom	Ikom (Ogoja)	4,708	4,516	47	849	Low
16.	Obrenyi	Obubra (Abakaliki)	3,249	2,692	30	424	Low
17.	Umuahia	Bende (Umuahia)	3,050	3,050	30	496	Low–Medium
18.	Obubra	Obubra (Abakaliki)	3,026	1,605	32	446	Low
19.	Arochuku	Bende (Umuahia)	2,580	2,434	30	444	Low–Medium
20.	Boje	Ikom (Ogoja)	2,310	2,021	24	240	Low
	TOTALS		25,583	19,832	234	3,365	

IV. *Cashew Estates*

21. Oghe	Udi (Enugu)	1,000	1,000	?	?	Medium

V. *Coconut Estates*

22. Bonny	Bonny (Degema)	970	863	?	?	Low
TOTALS		1,970	1,863	—	—	
GRAND TOTALS		148,931	67,072	815	11,945	

In each of the Corporation plantations, the exploitation of the land is based on modern land-use systems. The contiguous estate area is laid out in blocks devoted to regular rows of tree crops at the approved spacing. Internal roads, bridges, and embankments are constructed to aid movement of labourers, vehicles and machinery from one block to another for planting, cultivating and harvesting operations. The crops are expected to receive proper attention in terms of fertilising, mulching, weeding, pruning, and spraying with insecticides and fungicides.

Areas are set aside for the facilities associated with plantation operations: germination sheds and nurseries, buildings for housing and maintaining the processing machinery, offices, residential quarters and a recreational centre for supervisory staff and workers. Some of the estates have electrical generating plants, water-storage and pumping facilities, dispensaries, maternity homes and schools.

Many problems have emerged from the determined efforts of the Government to establish the ENDC plantations. Some arose out of ignorance or from errors of judgement concerning the physical and social conditions involved, others were probably unavoidable in the implementation of so ambitious a programme by enthusiastic but inexperienced and too few indigenous planners.

From the environmental viewpoint, the isolation of sites in inaccessible country hampered the field surveys necessary for the production of topographic, pedologic and cadastral maps. The field work was often carried out too hastily with subsequent errors in identification and in correct mapping of phenomena. Thus boundary disputes have persisted or have been provoked due to erroneous demarcation of the limits of the plantations, and crops have failed on soils incorrectly recorded as suitable for specific plants. The Ubani section of the Umuahia Cocoa Estate and much of the Arochuku and Obubra Cocoa Estates have failed to produce

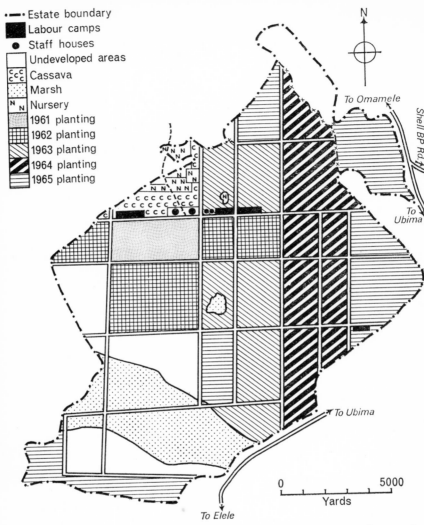

Legend

- •—• Estate boundary
- ■ Labour camps
- • Staff houses
- □ Undeveloped areas
- Cassava
- Marsh
- Nursery
- 1961 planting
- 1962 planting
- 1963 planting
- 1964 planting
- 1965 planting

To Omamele

Shell BP Rd

To Ubima

To Ubima

0   5000

Yards

To Elele

FIG. 12.3   *Layout of ENDC Rubber Estate, Elele (after Ogbonnaya)*

healthy trees despite the heavy costs involved in the initial establishment of the cocoa plants (Ubani is being replanted to coffee, Arochuku to oil palms). Marshy land or steep slopes within lands acquired for estates, unobserved during preliminary reconnaissance, have since come to light, sometimes reducing considerably the effective acreages which can be put to cash crops (Fig. 12.3). Elsewhere, the inclusion of areas of soils exhausted from many years of over-farming under primitive techniques means a costly rehabilitation programme before they can be utilised for commercial cropping. The isolation of many of the plantations has also involved heavy expenditure on road construction, transportation, and the importation of essential supplies.

From the social viewpoint, the complex intricacies of land-ownership patterns have on occasion not been fully unravelled, and legal agreements have been drawn up with villages who lacked authority to grant use of the entire land area being offered to the ENDC. Bitter disputes with other land-owning groups have resulted. Where attention has been given to the finer points of traditional tenure, estates of a bizarre shape have resulted – e.g. the Umuahia Cocoa Estate – which makes their operation as a viable entity that much more difficult (Fig. 12.4).

The remoteness of the estates has a depressing effect on the morale of the workers, who are removed from easy contact with kith and kin and the familiar, casual life of the village and its limited but accustomed amenities. The regular daily stint of manual labour also takes some getting used to after the less strenuous crop calendar of traditional agriculture. There is a dearth of adequately-trained supervisory staff and specialists for directing the operation. Those who join the services of the Corporation also find their posting to a distant area a real hardship.

There are innumerable technical and agronomic problems also, related to the raising of the tree crops, the combating of disease, and the processing of the harvested produce. In sum, the ENDC venture has thus far been a costly one, requiring completion of the accounts sheets in red ink each year. On the other hand, the financial success of the scheme can only be judged in the 1970s, when the present plantations will begin to yield revenue. It is highly doubtful, judging by the present employment figures (revealed in Table 12.3), that the optimistic targets of labour absorption over the Six-Year Plan period will ever be realised. These called for employment of 80,000 elementary school leavers, 15,000 school certificate holders and 2,000 university graduates.[1] At the same time, larger numbers of rural people may be encouraged to copy the work pattern and cropping techniques of the large-scale commercial plantations on a community level, utilising the basic 'nucleus' of the ENDC enterprises. Providing world prices are sustained for the three key tropical crops, oil palm,

---

[1] M. K. Mba, *The First Three Years*. A report of the Eastern Nigeria Six-Year Development Plan (Enugu, 1965), p. 9.

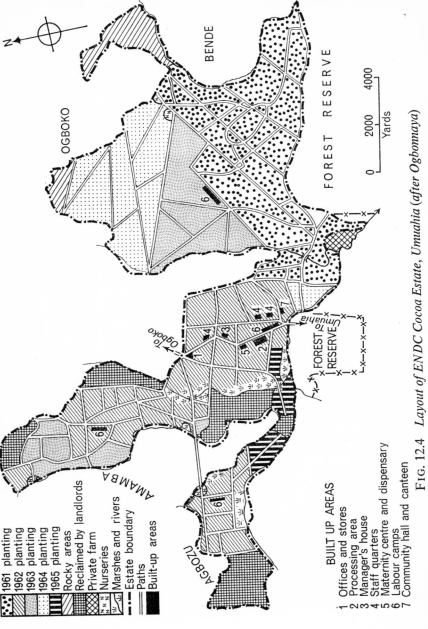

FIG. 12.4 *Layout of ENDC Cocoa Estate, Umuahia (after Ogbonnaya)*

OGBOKO

BENDE

FOREST RESERVE

0   2000   4000
Yards

To Ogboko

To Umuahia

FOREST RESERVE

AMAMBA

AGBOZU

1961 planting
1962 planting
1963 planting
1964 planting
1965 planting
Rocky areas
Reclaimed by landlords
Private farm
Nurseries
Marshes and rivers
Estate boundary
Paths
Built-up areas

BUILT UP AREAS
1 Offices and stores
2 Processing area
3 Manager's house
4 Staff quarters
5 Maternity centre and dispensary
6 Labour camps
7 Community hall and canteen

rubber and cocoa, the indigenous plantations of Eastern Nigeria should come into their own within the next decade.

(iii) THE FARM SETTLEMENTS[1]

The Farm Settlement Scheme is the most ambitious venture in the agricultural sector to revolutionise traditional farming systems and to teach Eastern Nigerians to produce cash crops through the application of modern agricultural methods. The farm settlements were allocated £6,125,000, or nearly 17 per cent of the total provision for primary production in the Development Plan. Under revised estimates of costs, the figure was raised in 1964 to £7,800,000 (21 per cent).

The farm settlements provide a practical demonstration of a new organisational approach to farming and rural settlement. They are state-organised and supervised co-operative farms modelled after the Israeli *Moshavim*, i.e., a plantation system of commercial agriculture in which the labour force, or settlers, instead of being merely paid workers as on a traditional plantation, have secure title on holdings of their own which they operate and from which they can draw an income, as well as having a share in the processing factory and in the running of the scheme, including the marketing of the produce. The farm settlements represent in fact co-operative ownership of expensive units of production, since they will be large enough to warrant installation of the same processing plant found in foreign-owned and ENDC plantations, and this machinery will belong to the Settlers' Co-operative Society. A settlement may also be viewed as a 'supervised credit' system in which the resident farmers are given substantial credit to develop and run an economic holding under supervision. Their farms and homes are acquired on hire-purchase agreements and the settlers are given some fifteen years to repay the capital from the time the farm settlement enters into production, six to seven years after establishment. The direction of the settlements by government personnel is scheduled to phase out eventually when the Farms are fully established and viable. The settlers themselves will be left to manage the community's affairs, including the processing mill and co-operative marketing of their products.

The scheme was launched in late 1961 and by the end of 1965 six farm settlements had been established, in six out of the twelve provinces of the

[1] For a fuller account of the farm settlements, see B. N. Floyd and M. Adinde, 'Farm Settlements in Eastern Nigeria: A Geographical Appraisal', *Economic Geography*, vol. xliii, No. 3 (July 1967), pp. 189–230.

H

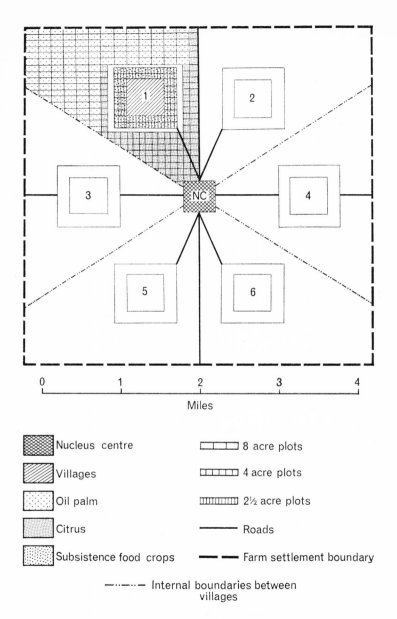

FIG. 12.5 *A model for a farm settlement*

TABLE 12.4

The farm settlements of Eastern Nigeria

| Name | Division and Province | Disposition of land (acres) | | | | | Cash crops | No. of villages planned | No. of settlers (mid-1965) |
		Total	Plantations and homelots	Residential, nucleus centre	Roads	Fuel plantations and 'waste' land			
Boki	Ogoja (Ogoja)	11,541	9,060	260	200	2,021	Oil Palm, Citrus	5	240
Erei	Afikpo/Enyong (Abakaliki/Uyo)	10,385	9,060	320	200	805	Oil Palm, Soya Beans	6	240
Igbariam	Onitsha (Onitsha)	6,560	5,915	324	120	201	Oil Palm, Citrus	3	350
Ohaji	Owerri (Owerri)	14,929	10,872	348	600	3,109	Oil Palm, Rubber	6	360
Ulonna	Bende/Okigwi (Umuahia/Owerri)	7,798	6,906	292	400	200	Oil Palm, Rubber	4	240
Uzo-Uwani	Nsukka (Enugu)	10,562	9,741	340	200	281	Irrigated Rice, Onions	5	95
TOTALS		61,775	51,554	1,884	1,720	6,617		29	1,525

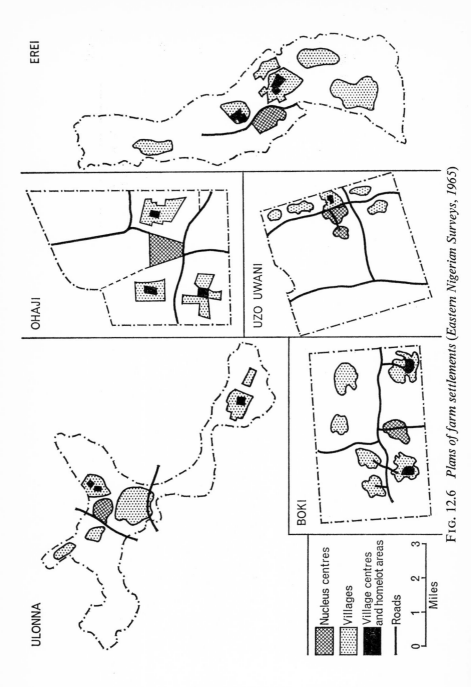

EREI

OHAJI

UZO UWANI

ULONNA

BOKI

Nucleus centres

Villages

Village centres
and homelot areas

Roads

0   1   2   3
Miles

FIG. 12.6 *Plans of farm settlements (Eastern Nigerian Surveys, 1965)*

Eastern Provinces. Each settlement is hopefully to comprise an area of some 8,000 to 12,000 acres. The settlers live in planned model villages of up to 120 families, and from three to six villages are scheduled for each settlement. Fig. 12.5 illustrates an idealised layout for a farm settlement. The villages consist of a grid street plan around which are located the settlers' houses, a community centre, a market, a junior primary school, the village square and playground, a communal yam barn and a firewood plantation. The villages are sited at a distance of about one mile from an administrative centre or settlement nucleus where all the services for the entire farm settlement are centralised. The nucleus centre will thus contain buildings for the processing mill, machinery pool and workshop, administrative and technical staff, the settlers' Co-operative Society, a market, health centre, police station, post office, churches and so forth.

The agricultural land is of two types. Large blocks of land suitable for plantation tree crops are subdivided into individual holdings of twelve to fourteen acres and managed under technical supervision. In addition to the settlers' share of the plantation acreages, each farmer has up to three acres of compound land for the production of subsistence food crops, specialised cash crops such as vegetables and pineapples, and the rearing of small livestock and poultry. Half an acre is in the vicinity of the homestead and the remaining acreage is located in the immediate environs of the village. The food crops are produced on an approved rotational pattern, under intensive cropping and fertilising techniques, and surpluses may be disposed of for cash in the markets of the settlement and beyond.

In sum, a total individual holding per settler family in a tree-crop farm settlement amounts to fifteen to seventeen acres. Settlers are not allowed to sublet or fragment their holding. All farmers are tenants-at-will to the Government of Nigeria, for a period of tenancy not exceeding thirty-five years, which is considered the length of the active working life of the first settlers.

Physical and human locational factors influencing the siting of the farm settlements are similar to those which have influenced the location of the ENDC plantations. However, a more determined effort was made to obtain suitable land close to the centre of heavy population so that the demonstration value of the farm settlements might produce the desired 'spread' effect on neighbouring communities, encouraging them to adopt (although on a more modest scale) the new land-use patterns of the farm settlements. Proponents of the settlements point out that the community plantations in fact partake of certain features of the farm settlement programme. However, the plan to locate the settlements within the high density areas has come only partly to fruition. None of the farm settlements is located within the main zone of densest population pressure since it was impossible to acquire the necessary minimum acreage. Rather do they lie on the flanks of the high density areas while, in the case of Erei and Boki, the settlements are located in lightly settled areas in the upper Cross River Basin. It may, of course, be argued that the location of the settlements on the boundary between areas of higher and lower densities will prove just as effective for demonstration purposes, with the added advantage that sufficient land is available nearby to enable communities to adopt the new methods of farming somewhat more easily than if they are in the heart of the high density zone.

Negotiations with the traditional landowners for the acquisition of land rights were more difficult and protracted than those required for obtaining sites for the ENDC plantations. 'Offers of land were often upset by the persistent oppositions of minority groups, in some cases threats to the lives of staff and surveyors, the meandering of traverses or sometimes complete abandonment of miles of traces for new ones.'[1] The claims and counter-claims of ownership of parts of the farm settlement land at Ohaji, Erei, Boki and particularly at Ulonna, led to lengthy delays before the land was eventually acquired by government. Worse still, they have resulted in an atrophying of size and a distortion of shape in half the settlements (Fig. 12.6).

The settlers are selected on the basis of their physique, age, educational standard, previous farming experience, and marital status. The original intention of recruiting school-leavers and other jobless youths (in an effort to relieve the acute problem of unemployment) has been dropped in favour of enrolment in the scheme of maturer, more experienced, married men with families, hence their own labour force. Persons who have worked on commercial plantations run by private companies, such as Dunlop Rubber or the ENDC, are particularly sought after, since they are likely to withstand the discipline and strangeness of early settlement life better than those who have scarcely left the familiarity and security of traditional village life before.

The ethnic origin of the settlers is also taken into consideration. Theoretically, the admixture of the farm settlement populations could be marked. However 'although people throughout Eastern Nigeria are free to enter into any Farm Settlement . . . certain restrictions [are] placed in their way in order to mitigate or eliminate sociological and political disturbances

[1] Eastern Nigeria, Ministry of Agriculture, *Annual Report, Agricultural Division 1962–63* (Enugu, 1964), p. 32.

and encourage as many of the original owners of the settlement land as are found suitable to be recruited as settlers'.[1] As a general guide, forty per cent of the settlers are chosen from the villages on whose land the farm settlement is sited; twenty per cent are chosen from the same division; another twenty per cent are selected from the same province; the remaining twenty per cent are recruited from other parts of Eastern Nigeria.

Igbariam Case Study. A good indication of the impact of farm settlements upon land-use patterns is provided by Igbariam, the first settlement to be established and the most advanced in 1965. Prior to acquisition the site was the scene of sporadic bush fallowing by tenant farmers from Enugu-Ukwu, a town about eight miles away from Igbariam, the principal land-owning community which lies to the north of the present farm settlement. The tenant farmers from Enugu-Ukwu had been on the site for over forty years and had built for themselves a number of semi-permanent dwellings. At the time of acquisition there were some forty settlement clusters or 'camps', each with an average of eight grass-thatched mud huts. Also on the site were a market; a school, used by the tenants' children as well as those from a neighbouring village outside the farm settlement area; and a maternity home of modest dimensions.

It has been estimated that between 1,000 and 1,500 people (including children) earned a livelihood from the land which will now be occupied, at fullest capacity, by 350 settlers together with their wives, their children and a number of adult relatives to make up their labour force. In other words, about the same number of people will henceforth be supported by the farm settlement on the same piece of land, although admittedly at a higher standard of living and with a demonstration of superior land use which it is hoped will strongly influence the surrounding villages. When artisans and shopkeepers eventually settle at the nucleus centre, the population of Igbariam may increase to around 2,000 inhabitants.

Under the tenants' land-use system, irregular-shaped and -sized pieces of land were being planted with annual crops, mainly cassava, for two to three years, followed by a fallow period of the same duration, with the result that the fertility of the soil was steadily decreasing. Only small areas of compound land were under continuous cultivation. Since the establishment of Igbariam Farm Settlement, the visible changes on the landscape have been marked. The patchwork pattern of scrub and bush fallow interspersed with islands of cultivation has been replaced by rows of citrus and

[1] Eastern Nigeria, Ministry of Agriculture, *Eastern Nigeria Farm Settlement Scheme* (Enugu, 1963), p. 9.

oil palms on the one hand, and by consolidated arable lands with neatly cultivated fields surrounding the newly-established villages on the other. In place of semi-permanent, primitive dwelling structures scattered almost at random through the bush, there are now regimented lines of modern cement-block and tin-roofed houses and other buildings associated with the service centres. The transformation is indeed impressive; however it is as well to remember that it has not been achieved simply or without considerable cost in terms of time, social disruption, a vast expenditure of human energy on the part of the planners and administrators and, not incidentally, of money.

Problems of Establishing Farm Settlements. Many of the technological and sociological problems which accompany the establishment of a farm settlement are common to those experienced by the ENDC plantation authorities. Considerable difficulties have been experienced at some of the farm settlements in the down-to-earth task of introducing settlers to their new life, avoiding at the same time undue psychological stress. The phenomenon of 'settler shock' manifests itself in a number of ways:

> The change of environment from the settlers' homes to the farm settlement sites imposes a lot of psychological strain on the settlers. To many of the settlers it will be the first time they must, by call of duty, leave the familiar environments of their own homes to go and seek a life on a farm settlement. Not only have they to be separated from their friends and relatives and the simple village routines they have known so well, but they are . . . taking a deep plunge into the unknown. Indeed the whole idea of the Farm Settlement is completely strange to them. For the first time many of the settlers have to conform to the new discipline of following a time-table of daily operations and observing rules of how to live and work together with their fellow settlers. . . . Amidst all these feelings of excitement, doubts and even depression at times, the settler is undergoing remarkable psychological strain which may manifest itself in the form of agitation, complaints, indifference, malingering, desertions, or feelings may run so high as to threaten the success of the farm settlement.[1]

Two aspects of the human geography of Eastern Nigeria which will

---

[1] J. C. U. Eme, 'Sociological Problems Connected with Farm Settlement Schemes', Appendix VI in Eastern Nigeria, Ministry of Agriculture, *Farm Settlement Scheme*, Third Annual Programme Planning Conference. Technical Bulletin No. 4 (Enugu, 1963), p. 61.

clearly not be solved by the Farm Settlement Scheme alone are those of unemployment and of chronic maldistribution of people in relation to the natural resources of the region. One of the original goals of the scheme was to relieve congestion in the most densely populated areas, and to create additional employment for an expanding population, thus stemming the flow of jobless school-leavers and others to the towns. The flow of primary-school-leavers in 1965 was running at nearly 100,000 young people a year. Of the boys, about ten per cent were going on to secondary schools; with luck, a further ten per cent found jobs in the new industries or as apprentices to various trades, but the great majority remained unemployed. Assuming that thirty per cent of the school leavers were girls (most of whom were shortly absorbed into the routines of home-making and motherhood), this still left an annual increase of 50,000 boys a year, to be added to a total of perhaps 350,000 jobless, partially-educated people in the region.

A maximum annual intake of 700 settlers a year, even if each of them has one or two school-leaver relatives whom he brings with him to his new farm, will not go far towards relieving this particular problem. Over the first three years, the settlements only absorbed 1,525 settler families. A maximum number of some 3,350 families by 1970 is again the proverbial 'drop in a bucket' compared with the employment needs of the Eastern Provinces. Modern scientific farming is clearly not a labour-intensive enterprise. Only in the hoped-for impact of the 'spread' effects of the settlements upon surrounding communities, and eventually more distant areas, can it be argued that the school-leaver problem is being softened by the Farm Settlement Scheme.

As for relieving congestion in the high density areas, the farm settlements are doing very little to demonstrate how more people can be encouraged to earn a living from a finite piece of land than are there at present; rather are they showing how a given community can earn a better living from the same piece of land which is farmed at present under traditional techniques. If, for example, in a zone where the density of population was in the order of 150 persons per square mile, the Igbariam Farm Settlement has at best maintained a similar density, what would be the impact of a farm settlement in the high density zone of Eastern Nigeria where, in parts of Owerri Province, rural densities are measured in terms of 1,000 to 1,400 people per square mile? One has to remember also that the tenant-squatter farmers have been obliged to forfeit their livelihood from the Igbariam area, and are now adding to the problems of rural over-population and under-employment in other parts of Onitsha Province.

Some, it is true, have moved to fresh under-populated areas in parts of Enugu and Abakaliki Provinces and are perpetuating their traditional squatter practices.

Perhaps the most serious criticisms of the Farm Settlement Scheme have been levelled by economists, who claim that the entire programme is far too costly for the results which it hopes to achieve. Some economists are in fundamental disagreement with the investment of so large a proportion of the first Development Plan budget in the agricultural sector, claiming that if such sums were invested in other sectors of the economy, more substantial and quicker dividends could be achieved, and in turn re-invested. There are others who regret that the Farm Settlement Scheme should occupy the premier position in the Government's endeavours to revolutionise agriculture, thus eclipsing (and financially crippling) a considerable number of less obtrusive but equally consequential measures for inducing the necessary technical and social changes in the rural areas.

For example it is claimed that, for a fraction of the excessive expenditure which the Farm Settlement Scheme is exacting, projects such as the Tree Crop Subsidy Scheme, fertiliser schemes, the extension system of small-scale demonstrations of improved farming and, above all, the community plantations, would achieve the same beneficial results in terms of a fundamental change in attitudes towards land tenure and land use, and more abundant harvests from the soil.

The time factor is also raised. Although in 1966, prior to the war, the planners spoke optimistically of a full planting-out by 1970, a decade may well pass before the farm settlements are entirely planted to tree crops, and over fifteen years before all the plantations come into full yielding stage. This period, it is argued, is too long a time for any scheme to attract foreign credit in the form of loans. Yet substantial aid in the formative years of any agricultural project helps to bring it to earlier fruition. Earlier returns from the labour and management inputs on a scheme the magnitude of the Farm Settlement Scheme would also help the settlers to develop a feeling for the financial possibilities of the project, thus sustaining their morale and confidence in the long-term success of the entire programme.

These are, to say the least, formidable arguments but the counter-arguments cannot be easily gainsaid. First, for those who are impatient for change, and reason that the farm settlements are proceeding too slowly for the expenditure involved, it is probably wise to remind critics of a well-tested axiom so far as large-scale planning in Africa is concerned: *festina lente* (make haste slowly). It is the intention of designers that the Farm Settlement Scheme shall be a going concern, a soundly-developed enterprise which shall stand in welcome contradistinction to the lengthy catalogue of ill-conceived and often abortive projects for agro-economic developments elsewhere in tropical Africa.

Even though the farm settlement project appears to be the favoured child of the Ministry of Agriculture, in strictly financial terms it is commanding (under revised estimates of costs) only 21 per cent or some £7,800,000 of the £36,821,000 allocated for primary production in the Six-Year Plan. The remaining 79 per cent is available for the many other programmes and approaches for achieving the desired transformation in the rural areas of Eastern Nigeria. It is not therefore a matter of either/or, but of combined efforts on many fronts, with a wide variety of schemes that are mutually beneficial and interdependent rather than competitive, in the all-out attack upon ignorance, poverty and malnutrition.

Protagonists for the farm settlements point out further that their project is a less expensive governmental method of achieving agricultural reforms than some of the other institutionalised programmes, in that almost all their costs will ultimately be repaid by the settlers. Costings of the establishment of a 12,000 acre settlement, with six villages, a large processing mill, the services of all government staff, personal emoluments and recurrent costs during the developmental period 1962-8 put the investment at £1,300,000. Of this sum about £1,000,000 will be debited to the settlers' loan, which will be paid off in fifteen annual instalments after the seventh year. The assumptions are that the farm settlements are run efficiently, that market commodity prices will remain firm, and that arrangements for reclamation of the loan are strictly enforced. Needless to say, these are big assumptions but, so far as commodity prices and the repayment of loans are concerned, the farm settlements are not unique in having to make them. The entire programme of innovations in the agricultural sector in Eastern Nigeria, in fact in all West Africa, is based upon similar provisos.

It is a salutary experience for economists to be reminded from time to time that the benefits to be derived from a particular project cannot always be measured in terms of direct cash returns alone. Any scheme which can capture the imagination of its participants and break through the crust of conservatism and suspicion would seem worthy of implementation. The farm settlements are an audacious, imaginative scheme to stir the rural populace into action, to create dissatisfaction at the old-established ways of earning a meagre livelihood from the land. They are furthermore a veritable showpiece on the landscape: a tangible, visible evidence of profound change in man/land relationships in the rural areas of Eastern

Nigeria. In this sense, they may pay far greater dividends than comparable sums of money absorbed in dribs and drabs in a thousand remote hamlets and isolated corners of the land: swallowed up in a sea of poverty with no visible evidence of the investment which might encourage others to seek a better way of life.

The sociological and psychological impact of such a programme as the Farm Settlement Scheme is immeasurable in direct monetary terms.

> One has to consider the degree to which a particular project provides a stimulus to economic activity generally and to estimate the extent to which it introduces, diffuses and popularizes new productive skills and new ways of economic life. One has to remember, too, that the 'critical minimum effort' required for take-off into sustained economic growth may demand a boldness in the initiation of projects which is incompatible with meticulous balance sheet calculations.[1]

'To give and not to count the cost' is a maxim which has at least some relevance in any broadly-based appraisal of the Farm Settlement Scheme. There are indications that external loan-granting agencies, noted for their strict concern over early viability of projects for which financial aid has been sought, are taking a new look at the farm settlement programme. In 1966 the Ministry of Overseas Development in the U.K. offered a loan of £1 million to help pay for the local costs of this important scheme.

Another economic consideration to be treated concerns the argument that an expansion in the commercial production of oil palms (the principal cash crop of the farm settlements) can be achieved considerably more cheaply by the tree crop subsidy programme, or by the establishment of 'orthodox' medium-sized commercial plantations.

The economist tends to overlook or ignore a number of problems which beset the Oil Palm Rehabilitation Scheme. These include the profound difficulty of supervising the planting and maintaining of new palms in a large number of widely scattered locations; the difficulty of harvesting, transporting and processing the fruit from such geographically-dispersed plots, the inexperience of junior extension staff and the reluctance of conservative traditional farmers to listen to their advice. Above all, there is the perennial problem of land tenure; few farmers can offer on their own the minimum of five acres which is required to qualify for a subsidy under the tree crop rehabilitation programme. Making up the five acres

---

[1] A. H. Hanson, 'Nile and Niger: Two Agricultural Projects', *Public Administration*, xxxviii (1960), p. 345.

by various means may land them in tedious and expensive litigation later on. Only the local chiefs or wealthy and influential persons in a community can offer five acres or more of land for replanting, and these are the people whom it is least necessary to help.

While such small-scale, 'grass-roots' innovations as the tree crop subsidy scheme may well prove worthy of their promotion in the long run, they can never take the place of large-scale demonstration centres such as the farm settlements, with the spectacular effects which they can engender on the whole fabric of conservative rural society, and with the model which they provide for a more radical transformation of traditional agriculture, so clearly required for the entire country.

So far as medium-scale oil palm plantations are concerned, it has been estimated that an economically viable 5,500 acre plantation, complete with processing mill, can be established for a net outlay of £200,000 over a ten-year period.[1] Such a plantation would earn annually some £340,000 at maturity and employ over 600 families, accommodating them under satisfactory community conditions. Nevertheless, the historic plantation system of agriculture still carries with it some of the odium attached to Western capitalistic exploitation of commercial agriculture in the former colonies of tropical Africa. Workers on plantations, as indicated earlier, are merely paid employees and have no personal stake in the land, in the running of the agricultural unit or in the marketing of its produce. Furthermore, such plantations do not lend themselves to widespread emulation among peasant farmers as do the novel and socially-stimulating community plantations, modelled in turn on the farm settlements, which point the way to a breakthrough to newer and more rewarding types of co-operative or group farming. The plantations provide little or no demonstration of new ways of producing staple foods for a rapidly expanding population: no illustration of spatial land-use planning for integrated rural communities of the future. It bears reiterating that the difficult-to-measure social dividends on capital investments in agricultural innovations are as important as the more easily recorded financial returns. The scrupulous concern of the economist for percentage profits must somehow be tempered with recognition of the non-quantitative attributes of improved agro-social relationships between husbandmen and their neighbours.

In sum, the objectives of the Farm Settlement Scheme appear both commendable and attainable. In face of the well-nigh Sisyphean task of

[1] D. L. MacFarlane and M. A. Oworen, 'Investment in Oil Palm Plantations in Nigeria: A Financial and Economic Appraisal,' Economic Development Institute, University of Nigeria (Unpublished Mss).

engendering changes for the better in the rural areas of Eastern Nigeria, the settlements are achieving undeniable advances. The ultimate aspiration of a fully transformed agricultural economy for the peasant population in the country is a worthy and noble one.

Whatever successes are accomplished will depend in the final analysis on the human element in the man/land equation; radical changes in land use and tenure cannot be implemented by legislation, paper plans or pipe dreams. It is the personal response from the farmers, both individual and collective, which will provide the key to success. Sustained enthusiasm and unmitigated hard work will ensure the attainment of the farm settlement exercise. A pithy and oft-encountered lorry slogan sums up the issue succinctly, in true Nigerian fashion: 'No Sweat No Sweet.' Which is, after all, merely a West African re-expression of a Biblical truism: 'By the fruits of your labours shall ye be known.'

13 Fisheries[1]

INTRODUCTION

FISHING is the last and most important of man's activities as a hunter. However skilled the pursuer or specialised the equipment, fishing is still the quest for wild creatures, free to wander as they will until caught in the net or on hooks. Fish-farming or pond-fish culture, on the other hand, is akin to livestock rearing. Fishing and agriculture are the two most significant primary occupations by which man attempts to produce sufficient foodstuffs to maintain an ever-growing populace. Just as there is an urgent need to introduce modern innovations into agriculture in Eastern Nigeria, there is a comparable need to transform traditional techniques of fishing among the region's population. Only by utilising improved fishing methods and new gear can Eastern Nigerian fishermen hope to make their contribution to the dietary needs of the region in the years ahead.

Serious and chronic malnutrition prevails in many parts of the Eastern Provinces. There is an absolute shortage of foodstuffs and the daily intake is unbalanced, heavily biased towards starchy foods (yam, cassava) and lacking in proteins and vitamins. In particular, there is a marked deficiency of animal protein. Only in the coastal areas, around inland rivers and lakes, and for favoured minorities of the urban population is the minimum daily requirement of animal protein satisfied. Crude estimates suggest that daily consumption of such protein amounts to seventy grammes per adult male along the coast, ten to twenty grammes in Owerri and Umuahia Provinces, and five grammes or less in Ogoja. The acceptable daily minimum in the U.S.A. is thirty grammes per adult male. These data emphasise both the relative deficits and surpluses of protein foods in broad geographical areas in Eastern Nigeria and the need for transporting fish products from the sea and rivers to inland populations.

Where obtainable, fish form a popular part of the diet of many Nigerians. The Eastern Nigerian population consumes at present an estimated twelve

[1] This chapter draws on a manuscript by J. E. Abaribe, 'Fisheries in Eastern Nigeria' (Nsukka, 1966). (Hereafter referred to as 'Fisheries'.)

pounds per head or 66,000 metric tons of fish a year. But of this figure, only 20,300 tons (31 per cent) are caught in Eastern Nigerian waters. Under normal conditions, a further 4,000 tons (6 per cent) are obtained from Northern Nigeria (Lake Chad), while by far the largest proportion, some 41,700 tons (63 per cent) have to be imported from foreign countries. Table 13.1 gives fuller details of the sources of fish consumed each year in Eastern Nigeria.

TABLE 13.1

Sources of fish supplies for Eastern Nigeria (1965 estimates)

Location	Metric Tons (Expressed as Fresh Fish Equivalents)
I. *Eastern Nigeria*	
1. Marine fishing (Continental Shelf, Gulf of Guinea)	11,800
2. Brackish water fishing (Creeks and lagoons of Delta and coastal lowlands)	6,000
3. Freshwater fishing (Rivers, lakes)	2,500
	20,300
II. *Northern Nigeria*	
4. Freshwater fishing (Lake Chad)	4,000
III. *Foreign Countries (Norway, Iceland, U.K. etc.)*	
5. Marine fishing (North Atlantic) Imported as:	
(*a*) Stock fish (dried cod etc.)	36,000
(*b*) Iced and frozen fish	5,000
(*c*) Tinned fish	700
	41,700
GRAND TOTAL	66,000

The characteristics and problems of fishing in the several different environments of Eastern Nigeria may now be reviewed.

MARINE FISHING

The continental shelf, or sea-bed to a depth of 100 fathoms, extends some thirty-five miles offshore from the Eastern Nigerian coast; but since small dugout canoes are used, fishing operations tend to remain close inshore (some two to three miles from land at most). The sea-fishermen are mostly Ibibio, Andoni and Ibo, with occasional clusters of Yoruba and even Ghanaians who have established settlements along the coast. The Ijaw have less keenness for sea-fishing and pursue their craft in the sheltered creeks, lagoons and distributaries of the Niger Delta. Recent surveys have recorded fifty-six fishing settlements east of the Bonny river extending to the Cameroon border; these communities operate some 1,600 canoes and have an average catch of 46·5 lb. per canoe per fishing day.[1]

The full list of fish taken off the Eastern Nigerian coast is lengthy and of little significance to anyone but an ichthyologist. Many of the species have only local, vernacular names. Some do not even have that: they are fish that just happened to get caught. But the main marine fishing is for *bonga* (*Ethmalosa sp.*) and sardines (*Sardinella sp.*). *Bonga* are found along the coastal banks of twelve fathoms depth or less, also near the river entrances and islands. Sardines are less tolerant of reduced salinity and are found in the open sea. Prawns are also taken along the coast or at the edge of the continental shelf.

Gear tends to be rudimentary: simple home-made basket traps, small throw nets, spears and hooks of bent wood and bone. For deeper water there are gill nets, cast nets and drift nets of different meshes. The major fishing hazard is the heavy surf which complicates launching and landing of canoes from the beach. The surf is caused by offshore or land breezes in the dry season, and the south-western 'monsoons' in the rainy season, which may create heavy swells or rough seas. Traditionally, fishing operations tend therefore to be restricted to the daylight hours and to the dry season. In any case, the catch of fish diminishes markedly during the rains. The yearly fishing cycle encourages men in coastal communities to be part-time farmers and fishermen – tending their land from May to October, fishing from November to April.

Clearly new gear and tactics are required if indigenous marine fishing is to make a fuller contribution to the Eastern Nigerian diet and economy. Larger, Ghanaian-type canoes, outboard motors and nylon nets would help the traditional fishermen, financed by credit from government-

[1] Abaribe, 'Fisheries', p. 23.

supported fishermen's co-operatives which would also provide instruction in modern fishing techniques and organise the marketing of fish by its members. Larger, more seaworthy vessels should also be brought into operation and experience gained in trawling techniques. Year-long trials with 36-foot and 50-foot Scottish trawlers have been undertaken and conditions are considered favourable for the creation of a commercial trawler fleet.

If mechanisation of canoe fleets, the use of improved gear and a trawler fleet were to materialise on any scale, processing, transporting and marketing of fish would need to be developed to meet the sharply-increased catches. Present techniques for attempting to preserve fish in the humid tropical climate of Eastern Nigeria are totally unacceptable as a basis for an improved fisheries industry.

> The usual subsistence fisherman's treatment of fish, while it prolongs their marketable life, does little to enhance their quality, taste or appearance. There can, in fact, be few less inviting sights and smells, than those of the open stands of blackened, fly-ridden and rotting fish found in some village markets of coastal Eastern Nigeria where production exceeds demand.[1]

The inefficiency of the traditional marketing structure should also be noted as a limiting factor in increasing protein consumption across the country. Middlemen often operate on margins which more than compensate them for the functions they perform. The close-knit organisation under which the too-numerous middlemen operate has given them a monopolistic control of the distribution of sea fish. This distributional system is sociologically deep-rooted in the economy and may not be easily altered.

In early 1966, however, a company – Fisheries (Nigeria) Ltd. – had plans to operate three eighty-foot steam trawlers, to be based on Amadi Creek at Port Harcourt. A cold-storage plant and ice-making facilities have been erected. Insulated lorries are available to transport the fish to inland consuming centres where demand is known to be great, e.g. Aba, Umuahia, Onitsha and Enugu. Local vendors are being recruited to distribute the fish more widely at fair prices, by means of small, ice-packed and insulated containers to fit on bicycles. Since 1963 a Nigerian concern – also at Port Harcourt – has been landing frozen fish caught in the trawling grounds to the west and south of Nigeria, by special arrangement with a foreign trawling company.

---

[1] Ibid., p. 18.

There are good prospects for tuna fishing off the Eastern Nigerian coast. However:

> tuna fishing operations demand large capital outlays for vessels and gear, for docking facilities, and for processing plants ashore. In addition, the more advanced fishing techniques are variable and highly specialized. Presently, the best approach in developing a tuna fishery would be to encourage foreign tuna producers to establish bases in Nigeria.[1]

Port Harcourt on the Bonny River might provide good bases for tuna fishing operations in Eastern Nigerian waters; so also might Calabar on the river of the same name, although it is off-centre with regard to domestic markets and still lacks good communication routes to the interior.

CREEK AND LAGOON FISHERIES

This sector comprises the complex network of estuaries, creeks and tidal lagoons which is the zone of mixing of sea- and fresh-water, the area of mangrove forest and coastal vegetation. It is developed at its best in the lower Niger Delta. Deep channels break into an extensive maze of shallow brackish streams which penetrate the mudflats of mangrove swamp and forest in jigsaw-puzzle fashion. Movement along the creeks is circuitous, but prospects for fishing never fail.

Practically everyone living in the Niger Delta, Ijaw men, women and children, engages in fishing, at least as a part-time occupation. With simple canoes of hollowed-out tree trunks, and various types of gear in the form of nets, hooks and traps, they catch mainly crabs, prawns, crayfish, catfish (*Clarias sp.*), *Tilapia*, sardines, oysters and *bonga*, which migrate into the creek areas at high tide from the marine sector. Some of the traps are ingenious and elaborate – cylindrical bamboo traps, ring and conical traps.

Because of the intensive subsistence exploitation of the fishing grounds in creeks and lagoons over the years, total production is low, averaging only about three lb. per man per day. Conservation measures probably need to be adopted, with restrictions on the number of units operating in these areas, whether they be individual fishermen in canoes, or groups in larger, motor-powered vessels. Certain sections should be closed to fishing, either seasonally or annually, to allow breeding and nursery areas

---

[1] I.C.A., *Fisheries Survey of Nigeria* (Lagos, 1961), p. 5.

for the fish (this rotation of fishing grounds is akin to rotation of paddocks for livestock farming). Elsewhere, the Ijaw should be encouraged to establish fish farms or ponds for the controlled rearing of fish for domestic consumption, even export, to the densely populated parts of Iboland to the east. An experimental brackish-water fish pond has been set up at Buguma.

FRESHWATER FISHING

All the rivers of Eastern Nigeria are fished but, except in a few localities, only for subsistence purposes, and invariably by part-time fishermen who are also farmers. In the freshwater swamp areas of the upper Niger Delta, and along the flood-plain sections of the Niger, Anambra and Cross Rivers, surplus quantities of perch, catfish, 'shineynose', *Tilapia*, prawn and other freshwater species may enter into local trade; but techniques for catching and preserving the fish are still rudimentary. Along sections of the Cross River, near Obubra for example, there are even riverain groups who have no fishing tradition. The Imo and Kwa Ibo rivers are fished from villages which have moved to their banks since the cessation of slave raiding. The larger lakes such as Ukwa and Oguta are also utilised.

The fishing craft are mainly paddle-propelled plank canoes and dug-outs. Spears and baskets are used in clear shallow pools during the dry season; other forms of gear include nets, traps, weirs and hooks. In the Niger below Onitsha light nets are used, attached to posts driven into the river bank or the side of the canoe. The gill net is popular on Oguta Lake, while hook-and-line techniques are practised by both adults and children almost universally. The bait ranges from *garri* and palm fruit to worms. Occasionally, stupefiers and poisons concocted from plants are added to streams and ponds.

Fish for sale are transported by canoe from one village to the next, while the versatile bicycle is used by hundreds of small hawkers who buy fresh fish at the waterside and pedal inland for distances of fifty miles or more to sell their wares. Excess fish are partially preserved by sun-drying, smoking or charring.

Even by introducing new equipment and techniques, it is unlikely that the freshwater catch from natural sources in Eastern Nigeria can be greatly enlarged. The most promising way of increasing the quantities of 'home-produced' fish is by artificially-stocked ponds – to be discussed shortly. Meanwhile by far the most important source of freshwater fish, albeit in preserved form, is Lake Chad in the far north-eastern corner of Northern Nigeria.

There are some 4,000 Nigerians fishing Lake Chad, producing annually about 15,000 tons of fish, 4,000 tons of which normally find their way to consumers in Eastern Nigeria. Fishing operations are traditionally conducted from papyrus or dugout canoes which are propelled by poling. Gear consists largely of lines of unbaited hooks (snagging) and cast nets. Improved methods of fishing using nylon gill nets and oar- or outboard-motor-driven fibreglass boats have recently been introduced.

All the fish, except those consumed locally, are air-dried or smoked. Ibo middlemen, who until recently had a dominant share of the trade in dried fish from Chad to the south, prepare the fish by cutting it into large pieces and smoke-drying it over wood fires (*banda*).[1] Despite this procedure, the fish are quickly infested with insects, particularly beetles; protein losses by the time the product reaches its markets in Enugu, Onitsha and Aba are estimated to be as high as forty per cent.

In view of the tedious and exacting journey from Lake Chad via camels or donkeys, thence lorries to Maiduguri, followed by the lengthy journey over poor roads to Eastern Nigeria, *banda* commands a fairly high price in the large urban areas of the south. With more efficient exploitation of Lake Chad's resources, new methods for salting and drying the fish to produce good keeping qualities and elimination of beetles, also improved roads and marketing arrangements, it is to be hoped that in the future greater quantities of Chad fish may reach Eastern Nigeria at more moderate prices.

POND-FISH CULTURE

A most promising way of increasing the output of edible freshwater fish is through fish farming, i.e. the construction of artificial ponds which are stocked with young fish or fingerlings; the fish are fed regularly and 'harvested' when mature.

The pond culture programme in Eastern Nigeria has not yet reached the magnitude of that in Western Nigeria, where a successful programme of communally-owned ponds at the village level has been in operation for some years. The first fish culture in the East was established at Umuna

[1] O. Nzekwu, 'Banda. The Secret of Ibo Concentration in Maiduguri', *Nigeria Magazine*, lxxix (1963), p. 248.

(Okigwi) in 1956; this now serves as the Government experiment station for testing different species such as *Tilapia spp.*, carp and catfish. It also functions as a hatchery for supplying fishpond owners with foundation stock free of charge. There are at present about seventeen ponds and reservoirs at Umuna, covering 115 acres.

The young fish are fed groundnut cake, soya beans, cassava, crushed beans, maize and rice, all feasible foods for farming communities to produce. Kitchen wastes and fertilising manures from the compound may also be used. There are problems of maintaining an adequate supply of fresh water and of checking invasion of ponds by frogs, also infestation by pests and parasites such as protozoa, fish lice, leeches and worms. But with a level of care and no greater supervision than that required for successful poultry farming, it has been shown that fish farming can be a profitable venture for co-operative groups in the rural areas, with an offtake of up to 1,100 lb. of fish per acre per year. With a more aggressive programme of publicity and demonstration, there is little doubt that pond-fish culture will spread rapidly. Some seventeen ponds were in existence in the region by the end of 1965.

A related possibility to fish ponds is the stocking of paddy rice fields with fish during the growing season, or raising fish as a second crop after the rice harvest. This type of fish culture has proved very successful in Java, Malaya and Japan, and there is no reason why it should not also succeed in Eastern Nigeria.

In sum, the consumption of fish from all sources – the sea, brackish- and freshwater natural locations, also man-made farms – could well be increased immeasurably in the years to come, with consequent benefits to the health and vitality of the inhabitants of the Eastern Provinces. Increased harvests from the waters can join with increased harvests from the soil to bring the region's economy to the much-desired take-off stage for sustained growth and rising prosperity for all its peoples.

14 Industrialisation in Eastern Nigeria

INTRODUCTION

If an agricultural 'revolution' is the goal of the Eastern Nigerian government, so also is an industrial 'revolution'. Substantial large-scale industrialisation along modern lines, proceeding hand-in-hand with rural reforms and the intensification of commercial agriculture, is urgently needed if the productivity of the region is to be increased and a marked improvement in living standards achieved for its inhabitants.

Diversification of the economy is a *sine qua non* for emergent countries. Subsistence or internal exchange agriculture, also reliance upon exporting a few primary commodities such as palm oil, rubber or cocoa, is no secure basis for the economy of an advanced society. Industrialisation helps to diversify exports and to increase foreign-exchange earnings and resulting national income. This diversification assists a country to withstand the shocks of fluctuations in world market prices for primary commodities. Import-substitute domestic industries can also help to check the mounting drain on national reserves which results from overseas spending on 'essential' consumer products and luxury goods.

The introduction of manufacturing and processing industries is equally important for the employment opportunities it creates, both directly in the factory and indirectly through ancillary services, transportation and marketing of the manufactured products. Experience in industry on the part of unskilled or semi-skilled rural migrants creates a reservoir of skilled labour which is then available for further growth and expansion of manufacturing enterprises. As the former Premier of Eastern Nigeria expressed it:

> Our industrial programme is designed to teach our young people the technical know-how and new skills with which we can make a technological breakthrough into the twentieth century.[1]

[1] M. I. Okpara, *The Purpose of Industrialization*, speech delivered at the official opening of Nigerian Glass Company Ltd., Port Harcourt, 24 August 1963 (Enugu, 1963), p. 4.

With massive, if disguised, unemployment in the rural areas, and the rapidly increasing number of school-leavers seeking jobs each year, alternative forms of employment in the urban areas of Eastern Nigeria are absolutely essential if rural rehabilitation schemes are to succeed and the region as a whole is to enter into the ranks of developed nations with high standards of living.

A further justification for industrialisation is the utilisation of untapped or under-exploited natural resources, whether organic or inorganic (vegetative or mineral) for increasing the gross domestic product. Adding value to previously-exported raw materials through processing or partial processing in Eastern Nigerian factories (e.g. palm kernels, rubber, crude petroleum) can also contribute to the economic advancement of the region.

A measure of the importance attached to industrialisation is afforded by the allocation of funds to industry and trade in the Development Plan, 1962–8. Some £12,930,000 or 17·2 per cent of the total capital expenditure was originally set aside for trade and industries. In the 1964 revised costs of the plan, this figure was reduced to £11,436,000 (15·2 per cent) but, after primary production (agriculture, forestry and fisheries), this allocation remains higher than that for any other sector.

TRADITIONAL INDUSTRIES

It is worth observing that a number of manufacturing skills in the strict sense of the word, i.e., the creating of useful products by hand, were practised within the area of Eastern Nigeria as far back as prehistoric times. Early bronze and iron articles, for example, have been unearthed at Awka and other localities. In more recent centuries, finds of iron slag near the Upper Coal Measure hills of the Nsukka-Udi Plateau attest to the smelting of iron for fashioning implements and weapons of war. Awka blacksmiths continue the traditions of iron-working, producing wrought-iron gates and a range of other metal wares to this day.

Other traditional handicrafts which have survived as 'cottage-type' industries include woodworking and carving at Awka, weaving at Akwete and Nsukka, pot-making at Inyi in Awgu Division, and raffiawork in the Ikot Ekpene and Itu areas. The products of these long-established industries reflect the efforts of skilled craftsmen and women, combining in many cases beauty of form and design with utility. The continued existence of craft activities is an important facet of the industrial scene.[1]

[1] T. A. Anumudu, 'Modernizing Our Local Crafts', *Trade and Industrial Bulletin*, v (Jan. 1966), p. 3.

As we have learned, the first major export from Eastern Nigeria after the cessation of the slave trade was palm oil. While this oil is relatively easy to produce, a manufacturing process of sorts is involved in its preparation for shipment overseas (as well as for local consumption). In the mid-nineteenth century some 30,000 tons of palm oil, valued at around £1,000,000, were exported annually from the Oil Rivers in wooden barrels. The oil was extracted from the palm fruit by hand- and foot-methods of expression, a partial manufacturing process still popular today in many villages.

In response to appeals from the Church Missionary Society, a number of agricultural processing appliances were sent to Nigeria as gifts, but since there was nobody on the Coast with adequate knowledge of their use, most of them rusted away unused or were broken through misuse. To remedy this situation, the CMS decided in 1852 to send out an Industrial Agent.

> This was almost certainly the first attempt to provide in Nigeria technical advice on industrial matters, but, unfortunately, Mr. Hensman died from an overdose of laudanum only four months after he arrived. He was followed in 1854 by Dr. Irving who worked for a year, assisting with industrial and commercial matters, before he died of dysentery.[1]

With the subsequent export of palm kernels as well as palm oil from the Rivers area, a need to replace the laborious manual extraction of the nut by machine power arose; in 1877 a piece of equipment was introduced which could produce some forty to fifty bags of kernels a day, thus doing the work of 600 labourers. This event may be said to mark the beginning of the Machine Age in Nigeria. However, the imposition of imperial authority at the turn of the century, and the economic policies of the British towards overseas possessions at that time, effectively barred the introduction of equipment on any large scale – machinery which might have led to the erection of manufacturing plants and processing of agricultural products or other raw materials being destined for the factories of Liverpool and other English towns.

Two world wars and a trade recession with world-wide ramifications prevented any major industrial development in Eastern Nigeria until around the middle of the present century. Exceptions to this statement are

---

[1] Anon, 'Industrial Development', *Nigerian Trade Journal*, Special Independence Issue 1960 (Lagos, 1960), p. 15.

few, the most significant being the initiation of the extractive industry of coal-mining at Enugu in 1917. Following the end of the Second World War, the need to improve the quantity and quality of oil palm products in the face of competition from other African and Asian countries encouraged the colonial authorities to approve the establishment of mechanised palm oil mills. The eventual construction of over one hundred Pioneer Oil Mills marks the beginning of the modern period of industrialisation in Eastern Nigeria.

LOCATION FACTORS IN MODERN INDUSTRIES[1]

Before describing and evaluating the existing pattern of both small-scale industries and large-scale establishments in Eastern Nigeria, we should review some of the major factors of location which influence the course of industrialisation in an underdeveloped country.

RAW MATERIALS AND POWER (FUEL)

These are fundamental requirements for modern industry. If the raw materials are more expensive to transport than the finished product, a factory saves on transport costs by locating as near as possible to the source of raw materials. Similarly if raw materials lose considerable weight in the course of manufacture, their processing industries are usually located near the raw materials. For example, in making iron and steel one needs several tons of raw materials (iron ore, ferrous metals, limestone, coke) to make one ton of steel. If one located an iron and steel plant close to the market or consuming centre, one would have to transport several tons of raw materials to market instead of one ton of iron and steel. It is therefore generally advantageous to locate an iron and steel industry near to the raw materials. (Fig. 4.2).

Another illustration of an industry which needs to be close to its source of raw materials is tropical lumbering, since up to fifty per cent of the felled trees or logs may prove unsuitable for the market after saw-milling operations. Hence the location of the Brandler and Rylke Cross River Mill near Obubra, close to the site of felling operations in the northern Calabar rainforest.

[1] This section is based on a manuscript, 'Industrial Location in relation to Nigerian Economic Development', by J. H. E. Johnson, formerly of the Geography Department University of Nigeria, Nsukka.

Where, on the other hand, a manufactured product *gains* in bulk or weight during processing, one saves in transport costs by manufacturing near the market. Furniture for example is more bulky than the sawn pieces of timber from which it is made, and thus the industry locates by markets (e.g. the Eastern Nigeria Construction and Furniture Co. Ltd. in Enugu). Another example of an industry which is best located near the consumer is brewing. The individual ingredients for making beer (barley, hops, water) are lighter than the ultimate product in crates or cartons of beer bottles and cans. The 'Star' Brewery at Aba and 'Golden Guinea' Brewery at Umuahia, both towns in the high axis population belt and on the railway between Port Harcourt and Enugu, are illustrations of satisfactory brewery locations.

If it is as cheap to transport the raw materials as it is to move the finished product, one might as well, other things being equal, take the raw materials for processing to a location near the market, or to a location where processing costs are cheaper due to power supplies: coal, oil, natural gas or electricity, or abundant labour.

In Eastern Nigeria, raw-material resources for manufacturing are abundant in some sectors, modest to meagre in others. Agricultural products provide the single most important source of raw materials for processing: oil palm, rubber, cocoa, fruits, vegetables, the staple root and cereal crops, tobacco and so forth. The high forests of Calabar contain many fine species of furniture woods. Mineral resources for industry include good deposits of limestone, moderate though dispersed quantities of low-grade iron ore, and limited deposits of lead and zinc ores, also salt.

Power or fuel resources in the Eastern Provinces are excellent, and more than adequate for a comprehensive industrialisation programme. The traditional resource of coal has been eclipsed since 1961 by the steeply rising output of oil and natural gas. The total output of crude oil has until recently been exported to Europe but the new refinery at Alesa-Eleme near Port Harcourt assures that domestic demands for petrol will be fully met in future. The road transportation requirements of industry will increase appreciably as new plants are established, and raw materials as well as finished products will require movement by petrol-powered vehicles. Natural gas is, among other things, being used to generate electricity. Since the exploitation and preparation of these mineral products for consumers constitute industrial activities in themselves, a fuller account of the coal and petroleum industries is given in Chapter 15.

LABOUR

Costs of labour are often a major item of the total manufacturing costs; where this is so, this is an important locational factor. In underdeveloped countries unskilled labour is usually plentiful and therefore cheap. On the other hand, skilled labour and managerial staff are in short supply and often expensive, especially where expatriate labour is concerned. Any industry requiring a high proportion of skilled staff is therefore an uneconomic venture in the early stages of industrialisation.

Unskilled labour for modern industry is abundant and cheap in Eastern Nigeria, thus representing an important resource. However, many of the young men seeking factory jobs in towns are quite new to industrial life and often inefficient in consequence. The Nigerian Petroleum Refining Co. Ltd. found it necessary to employ a number of Irish labourers during construction of the oil refinery at Alese-Eleme; Irish labour was also utilised at Tema in Ghana. The rural migrant in Eastern Nigeria may better first venture into the small-scaled industries such as leatherworking, tailoring and carpentry.

Needless to say, skilled workers are in short supply and expensive in Eastern Nigeria, as are managerial staff and entrepreneurs.

CAPITAL

Money with which to establish a factory is difficult to raise internally in a country where levels of income and domestic savings are low. Reliance upon external capital is therefore unavoidable and special incentives are necessary to attract private foreign investors.[1] In Eastern Nigeria these take the form of income-tax relief (or a 'tax holiday') for two years from the date when output starts, and liberal depreciation allowances on capital assets for 'pioneer' industries. A pioneer industry is defined as one which is not yet in existence in Eastern Nigeria or one which is not being conducted on a commercial scale suitable to the economic requirements for the development of the region. Efforts are also necessary to attract long-term, low-interest-rate ('soft') loans from governments of countries such as the U.S.A., U.K., West Germany, Italy, Israel and the Netherlands, or from organisations such as the International Bank of Reconstruction and Development (otherwise known as the World Bank), and the Inter-

[1] For some of the problems involved in attracting foreign industrial concerns to invest in Nigeria, see C. L. Mordi, 'The Problems and Planning for Industrial and Commercial Development in Nigeria', *Trade and Industrial Bulletin*, v (Feb. 1966), pp. 1, 3.

national Finance Corporation (an affiliate of the World Bank). At the national level, the Nigerian Industrial Development Bank is charged with attracting and dispensing capital for new industrial projects throughout the country.

Capital is required not only for the initial costs of erecting an industrial plant and installing modern machinery manufactured abroad, but for maintaining the machinery. This frequently means paying the salaries of skilled expatriate maintenance engineers, also importing spare parts from the overseas manufacturers.

TRANSPORT

The importance of satisfactory and inexpensive modes of transporting both raw materials and power to industrial sites, thence of shipping the finished products to market, is obvious. Access to harbour and railway facilities, or location upon good roads, are key considerations in the siting of new factories. The transportation systems (or the 'geography of circulation') in Eastern Nigeria are discussed in Chapter 17.

MARKETS FOR INDUSTRY

It is essential that a ready market either at home or abroad be available for the manufactured product. Ideally a market should be big enough (or show sufficient promise of growth in the years ahead) to enable an industry to operate at that level of output at which production costs are at their lowest. A large and expanding market assures 'economies of scale' leading to cheaper processing costs and more profits for the company. Clearly an emerging country such as Eastern Nigeria should not at present contemplate establishing industries for which the domestic market is known to be too small, e.g. the manufacture of television sets, transistor radios, cameras or washing-machines. If required, these 'luxury' products are best obtained from the advanced nations with highly specialised and efficient industries for their production, although their importation should be limited by means of high import duties during the formative years of an emerging nation's economy

With 12·4 million people, and a population which is still increasing rapidly, the *potential* market for manufactured goods in Eastern Nigeria is high. However the general poverty of the mass of the people at present means that the *effective* domestic market is small. Wage-earning and salaried

Easterners in the towns are obliged to spend an excessively high percentage of their income on foodstuffs, as indicated in Table 14.1.

TABLE 14.1

Percentage allocation of incomes on the part of urban populations in Eastern Nigeria[1]

Town	All items	Food and drink	Clothing and personal	Accomod- ation, fuel and light	Transport and other services	Other purchases	Tobacco Kola
Enugu	100	57	12	10	10	8	3
Port Harcourt/ Aba	100	51·5	11	15·5	10·5	9	2·5

GOVERNMENT INTERVENTION

This is a factor in industrial location which may affect any or all parts of the model. Yet it can be of paramount importance in making the ultimate decision as to where a factory is to be located, and whether or not it will operate profitably.

Government intervention may take the form of subsidising a new industry by providing a plot on an industrial estate at a nominal rent; paying for the training of indigenous labour overseas; or setting up a tariff 'wall' to prevent cheap imports of the same article from entering the home market, thus protecting the local factory and allowing it to charge a higher price for its products. Elsewhere, the Government may run factories itself, perhaps at a loss. The operations of the quasi-governmental Eastern Nigeria Development Corporation provide illustrations of this type of intervention. By 31 March 1963 the ENDC had accumulated losses since its inception in 1966 of nearly £1,300,000. Some £500,000 of this was on the Pioneer Oil Mills, £300,000 on the Oil Rivers Boat Yard and another £300,000 on the ENDC Plantations.[2]

There are numerous reasons for intervention by the Government. On economic grounds, it is feasible that a manufacturing enterprise may be

[1] Eastern Nigeria, Ministry of Economic Planning, Statistics Division, *Statistical Digest of Eastern Nigeria*, Edition 2. Official Document No. 24 of 1965 (Enugu, 1965), Table lii, p. 53. The figures are based on the results of enquiries into household budgets made in 1954 (Enugu) and 1957 (Port Harcourt/Aba), covering samples of labourers, artisans and clerical workers employed by government and commercial concerns whose basic earnings did not exceed £350 per annum (Enugu) or £400 per annum (Port Harcourt/Aba).

[2] ENDC, *Eighth Annual Report of the Eastern Nigeria Development Corporation 1962–1963, and the Accounts dated 31 March, 1963* (Enugu, 1963), Statement 1, pp. 35–36.

profitable in the future, even though it is making a loss today. An 'infant' industry of this type may not appeal to private sources of capital who want to see quick returns, and are unwilling to invest in the future. In such a case, if the industry is considered important enough in terms of the overall development plan for the Eastern Provinces, it may be worth-while for the Government to support the venture in its early years.

A further economic justification for a state-supported industry is that, while it may be unprofitable to its owners (i.e. the Government) it may be profitable for the community at large, offering employment in areas where before there was marked unemployment; it also provides a training situation for unskilled labour which, when it has acquired the necessary experience, may move on to other industries. In this way, the Government-financed factory produces 'external' economies for other industries. Thus, when private costs do not equal social costs, the Government may appropriately intervene to help or to run an industry.

On social or political grounds, Government decision-making may be motivated by national pride, a prestigous desire to demonstrate that one's country can manage a complex modern industry; lack of industries is often (and justifiably) associated with colonial status. Industrial development, being spectacular, is a tangible way of impressing the electorate that the Government is doing a good job and fulfilling its electoral promises to the people. Lastly, the location of a new factory may be decided upon by an influential politician, who wishes to see modern industries brought to his home area, thus satisfying his clansmen or local tribal union. This type of Government intervention prevailed at times in Eastern Nigeria prior to the military take-over of Government in January 1966.

15 Modern Industrial Developments[1]

INTRODUCTION

AN industrial enquiry carried out by the Government between 1961 and 1962 showed that some 671 industrial undertakings with ten or more employees existed in Eastern Nigeria at the time of the survey.[2] This figure includes both traditional handicraft centres, small indigenous industries and modern factory-type establishments. Nevertheless the survey resulted in an under-count, due to omission from the enquiry of handicraft and 'cottage-type' enterprises in the more remote areas of the country; the figure is certainly an underestimate today, since a number of new industries (large and small) have been set up as going concerns in the Eastern Provinces since 1962. If, furthermore, the 'cut-off' point had been taken as five employees or even less, a more realistic figure so far as small industries in Eastern Nigeria are concerned, then the number of industrial firms would have been increased to many thousands.

Assuming the existence of some 750 establishments of ten or more employees at the end of 1965, the majority of these are modest undertakings with limited capital investment and a labour force of from ten to fifty workers. Of the seventy sizeable or major industrial concerns with more than fifty employees (9·3 per cent of the total number of establishments), forty-one (5·5 per cent) have from fifty to two hundred workers, nineteen (2·5 per cent) have from two hundred to five hundred workers, and only ten (1·3 per cent) have more than five hundred employees (Table15.1).

[1] The reader is reminded that this chapter, as well as Chapters 16 and 17, dealing with industrialisation, urban centres and transportation systems, relates to the situation prior to the outbreak of the Civil War in 1967. The conflict has brought about considerable disruption and destruction of the pre-war patterns. However, it may be assumed that re-establishment of these industries and communication networks will occur during reconstruction, although certain locational and functional changes may be hoped for, and indeed are necessary if a more rational programme of industrialisation is to be mounted.

[2] Eastern Nigeria, Ministry of Economic Planning, Statistics Division, *Industrial Enquiry 1961–62*, Official Document No. 6 of 1964 (Enugu, 1964).

TABLE 15.1

Number of major industrial establishments in Eastern
Nigeria (1965, estimated)

Location	Province	Number of employees			Total
		50–200	*200–500*	*500–2000 +*	
Port Harcourt	P.H.	10	10	3	23
Eleme	,,	—	—	1	1
Enugu	Enugu	10	1	2	13
Emene	,,	2	—	1	3
Oghe	,,	1	—	—	1
Onitsha	Onitsha	4	3	—	7
Aba	Aba	5	2	—	7
Umuahia	Umuahia	3	2	—	5
Calabar	Calabar	2	1	1	4
Akamkpa	,,	—	—	1	1
Abonnema	Degema	2	—	—	2
Bonny	,,	1	—	—	1
Nkalagu	Abakaliki	—	—	1	1
Opopo	Uyo	1	—	—	1
TOTALS		41	19	10	70

SMALL-SCALE INDUSTRIES[1]

In every sizeable Eastern Nigerian community, from super-villages to the
periurban areas, from densely settled suburbs to the cores of the towns
themselves, there are clear indications through sight and sound of the
creative activities of small groups of artisans. Almost everywhere one finds
carpenters, tinsmiths, blacksmiths, wrought-iron workers, welders, tailors,
shoemakers and bakers. In the larger centres are printers, repairers of
typewriters and sewing-machines, vulcanisers, lorry- and bus-builders,
mechanics for repairing heavy equipment. The noise of the panel-beaters
and tinkers is very familiar in many places and surprisingly useful articles
such as storage trunks, kitchen equipment and lamps are fashioned from
second-hand sheets and tins. Other products of the small-industry sector
range from jewellery to plastic travelling bags, mattresses to nails and small
machines. Services include motor-vehicle maintenance, photography and
dry cleaning, also the repair of clocks, radios and bicycles.

[1] 'Small-scale' refers to the small number of people employed in an industrial firm or
workshop, e.g. a carpentry shop. It does not refer to the total number of people employed
in all carpentry shops throughout Eastern Nigeria. The total number of such carpenters
is in fact very large.

I

The emphasis and publicity directed towards large-scale factory-type industries, and the nationalistic pride which accompanies the opening of modern plants, have tended to overshadow the present role of small industry and its potential place in Eastern Nigeria's developing industrial economy. Yet it provides employment for approximately three times as many people as are engaged in large-scale industry, and supplies many of the goods and services enjoyed by domestic consumers.

> It acts as an incubator for the skills of the artisan and the manager. And most important, given the proper environment, small-scale enterprise is capable of greatly increasing its contribution to the Region's development. . . . Small industry is labour intensive: fixed investment – equipment and buildings but exclusive of land – is . . . about £100 per worker. Investment per worker in large-scale manufacturing is thirty times as great. Clearly an activity which makes intensive use of the Region's unemployed resource and economises on the scarce factor of capital is worthy of close attention and development support.[1]

In the aggregate, small industries may do more than large establishments in the immediate future to absorb the alarming number of jobless, partially-educated young men, estimated at 350,000 in 1965, and being added to by some 50,000 new school-leavers every year. The numerical importance of small industries in the fourteen towns surveyed by Kilby in 1961 is shown in Fig. 15.1, while the employment pattern is shown in Fig. 15.2.

The most significant feature of small-scale industrialisation in the Eastern Provinces is the almost staggering number of very small enterprises. In the fourteen towns examined in the survey, 10,728 firms were recorded, employing 28,721 workers. Thus the average-size firm employed 2·7 workers, including apprentices and the manager/owner. Seventy-five per cent of the enterprises were located in the four larger towns of Port Harcourt, Onitsha, Enugu and Aba.

It is in the area of workshops that the small industries effect their greatest capital saving. Over half (fifty-nine per cent) of the firms inspected by Kilby were housed in raffia sheds, corrugated-metal huts or market stalls. Cement structures or rooms in cement buildings constituted thirty-seven per cent of the workshops employed. The remaining firms worked in the open air. Forty-two per cent of the establishments possessed at least one machine, although only seven per cent had power-driven machinery.

There are naturally a number of problems to be overcome if small-scale

---

[1] P. Kilby, *Development of Small Industries in Eastern Nigeria* (Enugu, 1963), p. 1.

industries are to play a more important role in the region's economy. Technical management skills are low and capital is scarce.

A low level of *per capita* income, a surplus of labour and little technical knowledge are the salient features of the small industry landscape. . . . Rudimentary technical knowledge means that production techniques are simple, allowing practically anyone to establish a firm. Low levels of technical knowledge also mean that tools and machinery in use are seldom employed in the most productive manner.[1]

These problems may be overcome in part through government-sponsored technical training and management programmes for small-scale operators, together with capital loans for improved machinery and materials. Such a policy should lead to an increase in the quantity, quality and range of goods produced, also to reasonable prices for the consumer. Such assistance is now being provided at various training centres and through the loan services of the Fund for Agricultural and Industrial Development (FAID). Improved marketing techniques are also required in the areas of design, colour, packaging, labelling and advertising.

The prospects for the expansion of small industries may be assessed in an approximate fashion by examining the products which are imported into Eastern Nigeria, and by selecting those which have a simple technical process and can be produced on a small scale without sacrificing great economies of production. Such imported manufactured goods cover a very wide range, and include such items as household utensils of iron, steel, aluminium, earthenware, farm tools, nails, bolts, nuts, washers, screws and similar articles of base metals, a wide range of clothing including outer and underwear, umbrellas, pens, inks, paints and enamels, scented greases and pomades, insecticides, fungicides, foodstuffs such as biscuits, cakes, sweets and so forth. The value of these and similar items imported into Nigeria amounted to £16,874,000 in 1960, and this figure has been increasing yearly. The greatly expanded manufacture of import substitutes by small industries could make a significant contribution towards redressing the overall annual trade deficit as well as solving the long-term balance of payments problem in Eastern Nigeria. But, as indicated above, it is essential that their products be of good quality and competitively priced. In 1966 the Nigerian Ministry of Trade called on local manufacturers of goods:

to take the initiative to review their pricing policy so that products made in Nigeria cost less than similar goods imported into the country.

[1] Ibid., p. 5.

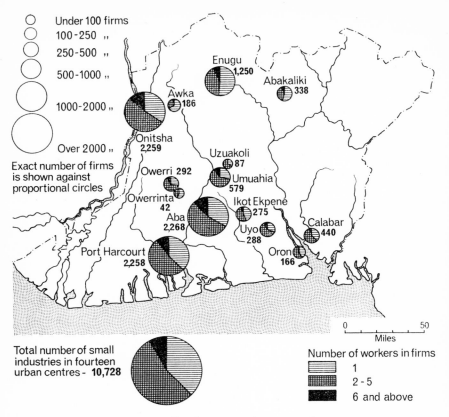

FIG. 15.1   *Small industries in fourteen urban centres*

Although the Government appreciates some of the manufacturers'
problems, it cannot continue to condone a situation where the
consumer is made to suffer because of patronising the products made
in his own country which are not competitively priced, and, in many
cases, are of inferior quality to those imported from other countries.[1]

LARGE-SCALE INDUSTRIES[2]

The major factory-type establishments in Eastern Nigeria may best be

[1] Barclays Bank DCO, *Overseas Review* (June 1966), p. 67.
[2] Acknowledgement is made of a working paper: 'Eastern Nigeria Industrial Mapping',
submitted by C. O. Muojindu, research assistant, Economic Development Institute,
University of Nigeria.

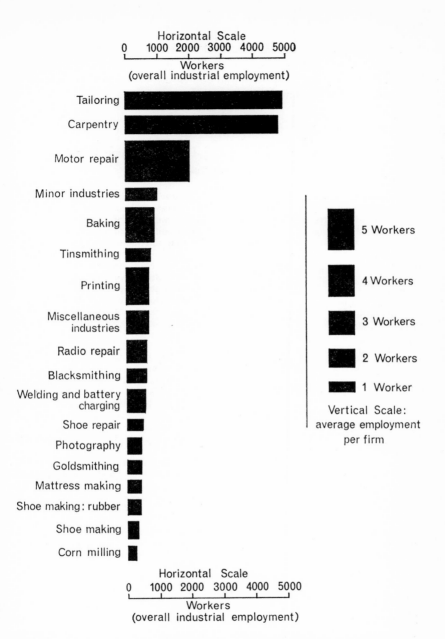

FIG. 15.2 *Employment pattern in small industries (after Kilby)*

listed in groups of related industrial activities, following – with some modifications – the International Standard Industrial Classification (Table 15.2). The origin, growth and contributions of selected 'case study' industries are reviewed in the following sections.

TABLE 15.2

Categories of major factory-type establishments in
Eastern Nigeria (1965)

I. *Extractive industries*
  1. Coal mining
  2. Drilling for petroleum and natural gas; oil refining
  3. Quarrying of stone, clay and sand

II. *Food, beverage and tobacco*
  4. Baking
  5. Brewing
  6. Production of carbonated waters and soft drinks
  7. Grain milling
  8. Processing of vegetable fats and oils
  9. Cashew nut processing
  10. Manufacture of cigarettes and snuff

III. *Textile, wearing apparel and footwear*
  11. Manufacture of textiles (spinning, weaving, finishing)
  12. Manufacture of clothing
  13. Manufacture of footwear

IV. *Tree, wood and paper products*
  14. Lumbering and sawmilling
  15. Rubber processing
  16. Manufacture of fibres (coir)
  17. Manufacture of furniture and fixtures
  18. Manufacture of veneer and plywood
  19. Printing and publishing
  20. Manufacture of stationery

V. *Chemicals, pharmaceuticals*
  21. Processing basic industrial chemicals
  22. Manufacture of drugs
  23. Manufacture of soap
  24. Manufacture of cosmetics

VI. *Glass, clay, cement products*
  25. Manufacture of glass
  26. Manufacture of ceramics
  27. Manufacture of drilling clays
  28. Manufacture of cement
  29. Manufacture of asbestos cement products

VII. *Metal products*
  30. Aluminium: rolling mill
  31. Iron and steel: rolling mill
  32. Manufacture of galvanised and wrought-iron products
  33. Manufacture of metal window and door frames
  34. Manufacture of vitreous enamelware

VIII. *Machinery and transport equipment*
  35. Lorry-building and vehicle repairing
  36. Tyre manufacturing
  37. Assembly of bicycles
  38. Boat-building and repairing

IX. *Miscellaneous*
  39. Manufacture of gramophone records
  40. Manufacture of plastic products

(I) COAL-MINING

As mentioned earlier, coal-mining is the oldest industry along modern lines in Eastern Nigeria. Production of coal began in 1917, following completion of the railroad from Port Harcourt, and in the peak output year, 1958, five mines were operating: the Iva Mine, Ogbete Mine, Okpara (Hayes) Mine, Ekulu Mine and Ribadu Mine.

The collieries at Enugu, managed by the Nigerian Coal Corporation, are the only ones in Nigeria, and indeed in all West Africa. The principal customers of the corporation have traditionally been the Nigerian and Ghanaian railways, also the Electricity Corporation of Nigeria. A twenty-five-mile aerial cableway carries coal from Enugu to the ECN thermal power station at Oji River. Due to programmes of dieselisation adopted by both national transport systems, also increased reliance upon oil, natural gas and hydroelectric power for the national grid of electrical power, as well as high production costs for the coal itself, the industry has been in a distressed condition for some years with 1963 production figures on a level with those of 1947, sixteen years previously.

The hope for the future lies both in streamlining and improving the efficiency of the mines and in seeking alternative uses for Enugu coal. The industry will be viable and a boon to industrial development in the region, rather than a drag, only if costs and prices can be cut considerably, with the latter at a genuinely competitive level of those of other fuels now available in Eastern Nigeria. Output per man shift has in fact more than doubled in recent years due to the adoption of new operating techniques.

The coal itself is of variable quality only: generally sub-bituminous. Tests have indicated that coke can probably be derived from Enugu coal (by yet insufficiently-developed processes) but it is doubtful whether such coke is metallurgically suitable for blast-furnace work for an iron and steel industry. Alternative uses being investigated include the production of synthetic fertiliser (sulphate of ammonia), tar fractions, light oils, benzole and other chemicals. There is enough coal in the vicinity of Enugu to last a further fifty years at the present rate of exploitation. Larger reserves have been located in Northern Nigeria.[1]

(ii) PETROLEUM AND NATURAL GAS

The largest industrial concern in Eastern Nigeria, if not the entire Republic, is the Shell–BP Petroleum Development Corporation of Nigeria Ltd., engaged in oil exploration and production. From humble beginnings in 1937, when Shell D'Arcy interests were granted authority to engage in geological surveying and mapping, the company had grown by 1966 into a giant organisation with approximately 4,000 direct employees, a capital

[1] Nigeria, Geological Survey, *The Coal Resources of Nigeria*, Bulletin No. 28 (1963).

investment of over £150 million and an annual output of 15 million tons of oil (from Eastern and Mid-Western Nigeria). Nigeria had become the second most important producer of oil in the Commonwealth, after Canada, and crude oil was Nigeria's principal export, exceeding in importance the traditional agricultural exports of groundnuts, cocoa and oil palm products. Valued at £68 million in 1965, ninety per cent of this impressive export of oil originated from Shell–BP.

The early explorations were interrupted by the Second World War but resumed in 1946 when the tempo of work was greatly increased, aided by new techniques in seismic geophysical surveying and aerial photography, which accelerated reconnaissance mapping. Negotiating the tropical forest and swamps of the Delta on the ground for purposes of mapping is no simple matter. The arduous work of exploration is also aggravated by heavy rainfall for many months of the year and widespread flooding by the Niger in its lower mangrove swamp reaches. The difficulties presented by terrain and climate are even more pronounced when the development stage is reached, and heavy derricks, rigs and other equipment must be brought to promising sites for test drilling.

The drilling of exploration wells commenced in 1952 but, apart from a trace of oil at Akata in Uyo Province (drilled in 1953–4), a succession of dry wells were sunk, and the situation was most disheartening. At last, in January 1956, oil was found in commercial quantities at Oloibiri in Yenagoa Province, forty-five miles west of Port Harcourt (Fig. 15.3). Nineteen years had passed since the search began and £15 million had been expended.

In the hope of placing Nigeria on the world oil map, Shell–BP decided to develop the Oloibiri field, although the task of producing oil in the swamps and transporting it to the coast for shipment overseas was a formidable one. A sixty-mile pipeline to Port Harcourt was constructed, at a cost of over £1 million. Some sixteen appraisal wells were drilled, but only eleven proved productive and the field as a whole failed to live up to expectations; although pumped for a few years, production fell off steadily – by 1962 only two wells were yielding oil – and water entered in increasing quantities.

At the end of 1956 oil was found in sufficient quantities for commercial exploitation at Ebubu and Afam in Umuahia Province, about twenty miles east of Port Harcourt. Other discoveries followed in quick succession in this new zone, Bomu in Ogoni Division being an important location with many producing wells. An important event took place in February 1958 when the first shipment of Nigerian oil left Port Harcourt for a refinery in Europe. A total of only 245,000 tons of crude oil was exported in that

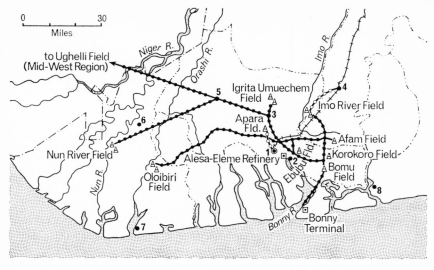

Oil fields
- ▲ Oil fields
- ●━●━● Oil pipelines
- ○━○━○ Natural gas pipelines
- ◙ Shell-B.P. installations (refinery and terminal)

1 Port Harcourt	**5** Rumuekpe
2 Okrika	**6** Yenagoa
3 Agbada	**7** Brass
4 Aba	**8** Opobo

FIG. 15.3   *Shell–BP oil-producing centres*

year, a mere 1·6 per cent of the 1966 production figure eight years later: but it was a beginning at least. Additional pipelines linking Afam and Bomu with Port Harcourt were laid during 1958 and in 1959 considerable progress was made, with the drilling of thirty-five exploratory wells, nine of which discovered oil, and sixteen appraisal wells, all of which were productive. Natural gas was found in association with the oilfields but since only a very small portion could be used economically, most of it had to be flared off.

Work began on an ocean tanker terminus on Bonny Island at the mouth of the Bonny River, some thirty-five miles downriver from Port Harcourt, to be connected by a new pipeline from Bomu, Afam and the other producing fields. A bar across the entrance to the river was dredged experimentally, to ascertain whether a twenty-nine-foot channel could be maintained at a reasonable cost to allow large tankers to go in and out of the Bonny terminal. The terminal, with a storage capacity of 66,000 tons, was completed in 1961 and oil was piped to ships berthed at buoy moorings in the Bonny River. The tankers were small, since 18,000-ton vessels could only clear the bar when half-loaded; a 12,000-ton 'shuttle' tanker had to be maintained in the river to 'top-up' the larger tankers through a submarine line at tanker moorings eight miles out to sea. This was an expensive and time-consuming operation which materially raised production costs of Nigerian oil and made it uncompetitive on the world market.

Since 1961, when the oil began to move overseas in business-like quantities, Shell–BP has grown from strength to strength. In 1962 further fields were opened up at Imo River and Igrita Umuechem, a flow station was constructed at Bomu, and work began on doubling the capacity of the Bonny terminal. Plans to utilise the important resources of gas led to construction of pipelines from Apara to the Trans-Amadi Industrial Estate at Port Harcourt, and to the site of a proposed gas turbine electricity-generating plant at Afam. Shell–BP also acquired exploration rights to four of the twelve offshore concessions of 1,000 square miles each which the Federal Government had demarcated along the continental shelf.

With the success of the Shell–BP operations other groups began to acquire both onshore and offshore oil-prospecting licences, among them Nigerian Gulf Oil Company, Mobile Exploration Nigeria, American Overseas Petroleum Ltd., Tennessee Nigeria Inc., AGIP, Mineraria and Safrep (Nigeria) Ltd. By 1966, some eight companies representing American, British, Dutch, French and Italian interests were actively engaged in the exploration and production of oil in southern Nigeria.

In 1963 new discoveries in the Delta proper, the Nun River (Yenagoa) field, also at Korokoro (north of Bomu), were made by Shell–BP; nine producing fields came into operation, supported by nine flow stations, and experimental offshore drilling commenced. Africa's first gas power station, built at a cost of £2,500,000 opened in May 1963. The Afam gas-fired turbine power station produces electricity at substantially cheaper rates than conventionally-operated electrical power stations; electricity from Afam is supplied to many towns in the southern part of Eastern Nigeria. The export of crude oil was further facilitated by the dredging of a twenty-seven-and-a-half-feet deep channel across the Bonny Bar and by the end of 1963 the largest single shipment of 26,000 tons was evacuated. This record was to be broken repeatedly in subsequent years, e.g. 37,400 and 47,000 tons in 1964 (the channel now being thirty-five feet deep, due to dredging operations by the Nigerian Ports Authority), and 65,750 tons in 1965; more than one-quarter of the entire first year's output in 1958, in a single vessel.

In 1964, Shell–BP had six drilling 'strings', or teams, in action, including offshore activities some twenty miles east of Bonny. The terminus was equipped with eight storage tanks and the oil was moving smoothly and efficiently from wells to export tankers. An important step towards increased domestic utilisation of the power resources was the completion of an eighteen-and-a-half-mile aluminium gas line from the Imo River field to the Industrial Estate at Aba.

A far more significant and lengthy system, the Trans-Niger Pipeline, was opened in 1965, connecting the newly-productive Mid-Western oil fields to the evacuation facilities at Bonny (Fig. 15.3). The pipeline was a major engineering feat, traversing rivers, swamps, thick forests and farm-lands for a distance of 140 miles, and at a cost of over £5 million. Also in late 1965 the most important facility so far as the internal needs of the country are concerned – a Nigerian oil refinery – went 'on stream' at Alesa-Eleme, twelve miles south-east of Port Harcourt.

A Nigerian oil refinery had been mooted as far back as 1954, when under the terms of the oil-mining lease granted by the Federal Government to Shell–BP the Company pledged that they would consider the economic feasibility of establishing a refinery, as soon as the output of crude oil attained an aggregate of 500,000 tons a year. With the target figure reached in 1959, a survey team set about selecting a suitable site for the refinery.

> Their task was not an easy one. Six alternative sites were eventually selected, three of them in the Lagos area and the remainder around Port Harcourt, these being the two principal ports along the Nigerian coast. The team were looking for about 300 acres of land, firm enough to support the very heavy and expensive plant to be erected on it, close to an existing road and railway network, within easy reach of deep water and the open sea, and not too far removed from the sources of crude oil. After due consideration of all these factors, and in particular the cost of sea and inland transportation of products, together with other problems such as salinity in the atmosphere with its corrosive effect upon the plant and equipment, it was finally decided that the site most nearly meeting all requirements was at Alesa-Eleme. But, like all projects of this magnitude, no site is perfect and the chosen area and pipeline route to the Bonny River had their own special problems, all of which had to be overcome.[1]

Most of the three hundred acres are taken up by the tank 'farm' where crude oil, partially-processed products and finished products are stored. Since supplies are received direct from the Shell–BP producing fields in the

[1] Anon, 'The Nigerian Petroleum Refinery Project', *Trade and Industrial Bulletin*, vol. iv, No. 40 (Nov. 1965), p. 1.

vicinity, there is less need for crude storage than there would be at a refinery dependent upon deliveries by sea. The oil refinery plant itself – crude distillation unit and a catalytic reformer – has attached to it in addition a special plant for the production of liquefied petroleum gas.

To connect the refinery to a marine loading jetty on Okrika Island, whence products may be transported to market by tankers on the Bonny River, a ten-line pipe track had to be laid down for a distance of three miles through a tidal swamp and a large area of mangrove forest. This entailed the construction of a trestled structure over 1,000 yards in length supported on piles sunk ninety feet deep into the mud. The structure is designed to allow sufficient clearance for the canoes of traditional Okrika fishermen to pass under it.

The Alesa-Eleme refinery of the Nigerian Petroleum Refining Co. Ltd. was constructed at a cost of £10 million, £4 million of which was spent on local appropriations. It will use about 1·9 million tons of Nigerian crude oil annually, to meet total domestic demands for motor spirit, kerosene, gas/diesel oils, fuel oils etc., and to yield an additional 460,000 tons of petroleum products for export. The whole operation will result in tremendous savings to Nigeria in foreign currency, helping to redress her imbalance of trade and releasing funds for investment in other development programmes.

> The establishment of the refinery can therefore be aptly described as a corner-stone in the structure of industrial development in Nigeria. Its growth and development will be insulated from external influences since it depends on the oil fields in Nigeria for its crude oil. It will also provide security for Nigeria both from defence and economic angles in the event of an emergency outside the country.[1]

The Shell–BP venture in Eastern Nigeria has thus paid off the most handsome dividends to the economic development of the region, and is destined to play an even greater role in the years ahead. Prior to the outbreak of the Civil War in 1967, the expansion rate was unchecked: fourteen storage tanks at Bonny in 1966, eight drilling rigs, seven seismic parties, additions to feeder pipelines, more roads and bridges, and an expansion of the important technical training programme for Nigerians. The seven other oil companies were also noticeably stepping up their activities. The significance of this investment in men and equipment is incalculable in terms of the future prosperity of the Republic.

[1] Anon, 'Nigeria's Oil Refinery', *Nigerian Trade Journal*, vol. xi, No. 4 (Oct./Dec. 1963), p. 154.

In early 1967 Nigeria was the third most important producer of oil in Africa (after Algeria and Libya). While her output in 1975 will be scarcely 1·5 per cent of the estimated world production of 2,000 million tons in that year, nevertheless in terms of foreign exchange earnings and its pride of place in the internal economy, the oil industry may well provide the much sought-after panacea to the underprivileged status of so many Nigerian citizens.

(iii) BREWING

Brewing ranks among the important industries of Eastern Nigeria since it is a large employer of labour and, through a diversity of jobs, provides opportunities for acquiring skills and technical knowledge vital to the development of the region. It yields sizeable annual revenues to the Government in tax, also in customs and excise duties, and has led to an appreciable amount of saving in foreign exchange.

There are two breweries in the Eastern Provinces. The Star brewery at Aba began production in 1957, and is owned by Nigerian Breweries Ltd.

> The initiative to form the Nigerian Breweries Limited came from the United Africa Company Limited who, with other merchant firms, invited the internationally known Dutch brewers, Heineken's, to join them as partners in the enterprise. U.A.C., with its long experience of business in Nigeria, became managing agents. Heineken's provided the technical expertise to ensure from the earliest days that the local product would always be of the highest quality.[1]

Aba was selected as a suitable site since the town is a centre of road and rail communications which are important not only for outward distribution but also for inward return of bottles. The market for beer is region-wide but highest sales coincide, not unnaturally, with the area of high population density and the urban centres. In the distribution of its products[2] the aim of Nigerian Breweries has been to preserve price stability throughout the Eastern Provinces. By delivering free to all the main centres in the region, the company ensures that its products are bought at the same price up-country or in the bush as in Aba. Field managers travel around taking orders, and the goods are transported by outside contractors. The geo-

[1] Anon, 'Breweries in Nigeria', *Nigerian Trade Journal*, vol. xii, No. 1 (Jan./March 1964), p. 10.
[2] In addition to beer, Nigerian Breweries produce a range of mineral waters and Schweppes products.

graphical factor of distance from producer to consumer is thus deliberately countered by means of differential costings.

In 1963, the Golden Guinea Brewery at Umuahia was opened, at a cost of £1 million. This plant was the first indigenous brewery in Nigeria, financed by the Eastern Nigerian Government (ENDC) as part of the many industrial ventures planned under the Six-Year Development Programme. The brewery was built by the Eastern Nigeria Construction and Furniture Company and the machinery was installed by a German firm. A number of German experts are employed at the brewery to teach Nigerians modern methods of brewing, but the management of the Independence Brewery Ltd. is in the hands of Nigerians. The locational factors relating to Aba are equally applicable to Umuahia.

Both breweries are capable of producing about 1·5 million gallons of beer a year; their combined output amounts to about two-thirds of the beer consumed in Eastern Nigeria. Both plants, as well as the Pepsi-Cola Bottling Plant in Onitsha, are supplied with bottles from another ENDC-sponsored enterprise, the £1·2 million Glass Factory in Port Harcourt, with an annual capacity of some 40 million glass containers or bottles a year.

(iv) LUMBERING AND SAW MILLING

Nigeria is among the world's most important producers of tropical timbers, although the bulk of her timber production and exports have traditionally come from saw-milling establishments west of the Niger River. The industry has also been directed towards overseas markets, although local consumption is now increasing at a rapid rate. Some eighty-five per cent of Nigeria's timber exports is in the form of logs, only fifteen per cent comprising sawn timber and plywood. The value to be added to wood exports through further domestic processing warrants the erection of additional factories for plywood, chipboard, etc.

There are over one hundred usable timber species in the high forests of Eastern Nigeria but less than twenty – the so-called commercial species – have been introduced to the world market. The lesser-known species, though often of comparable quality to the commercial trees, are seldom felled or extracted in large-scale lumbering operations since their versatility is not fully appreciated in the European or American furniture trade. Yet Eastern Nigeria has a wide range of timber species suited for all purposes.

There are, for example, timbers light as Balsa – *Ricinodendron*; some extremely heavy – *Ekki*; some as white as sycamore – *Funtumia*; some as black as soot – *Ebony*; some crimson red – *Camwood*; some as silky as Polar Birch – *Celtis*; some more durable than iron – *Erun, Ekki*; some with a very fragrant smell – *Scented Guarea*, and others with a rather pungent smell – *Cylicodiscus*.[1]

Apart from furniture and interior decorating, there are species which may be used in boat-building, for railway sleepers, pulp and paper manufacture, matches, charcoal, carvings, sports goods and pharmaceutical products. 'In fact, it can be modestly claimed that at least one Nigerian timber is available for every conceivable use to which timber is put.'[2]

Before investing money in a sawmill, heavy equipment, vehicles, staff housing, approach roads and so forth, a lumbering company must be assured of long-term timber concessions so that the source of raw material can be relied upon. Agreement must be reached between local authorities and the Government on fees to be paid and the royalties on all trees felled and extracted.

In 1956 the Eastern Nigerian Government invited the expatriate firm of Brandler and Rylke to take up timber concessions in the upper Cross River Basin near Obubra. Before that date the reserve forests in the area lay largely unexploited except for a few pit-sawyers extracting timber and indigenous owners collecting firewood. Villages such as Apiapum were primitive and isolated. The laterite road linking the old Government station and Divisional headquarters, Obubra, with Ediba and Abakaliki, the Provincial headquarters, was the only motorable road in the vicinity.

Brandler and Rylke took up the proffered concession in the 320-square-mile Cross River forest reserve and commenced operations in 1957 after sizeable investments in roads, a landing beach, and the Cross River Sawmill, the largest in Eastern Nigeria. Logging is the main activity of the concessionaires. The process commences with a thorough survey of the forest area in order to assess its worth. A square mile is divided into eight-acre blocks in which the position and size of the various species of exploitable timber are plotted. As yet, only an average of one tree an acre is useful; such a scattered distribution means that the expensive network of extraction roads built into the previously inaccessible forest must be expertly laid out.

The extraction operation starts with the felling of selected trees by axemen; after felling the trunk of the tree is hauled out by a powerful tractor to a clearing known as the 'landing ground' or gantry. Here the

---

[1] Nigeria, Federal Ministry of Information, *Some Nigerian Woods* (Lagos, 1962), p. 4.
[2] Ibid.

boles are collected together, stripped of their bark and cross-cut into suitable lengths. The logs are then loaded on to massive timber lorries for their journey to the beach on the riverbank or to the sawmill. The rafted logs, or 'floaters', move downriver to Calabar for export during the rainy season, an unfortunately short operating period from May to September. *Obeche* logs invariably go for export while species such as *Agba, Indigbo, Iroko, Apa, Red Camwood, Ironwood, Limba* and *Opepe* go to the mill.

> The sawmill logs are placed on the band-mill carriage which rolls backwards and forwards past a vertical band saw known as the headrig. At every forward movement, slabs of wood are cut from the logs. Powered rams and arms turn the heavy timber round to present the best face to the saw. Operating such a machine calls for great knowledge and speedy judgement. The slabs fall from the log direct on to a powered transfer system which moves them to other saws – edging, trimming, and further cutting them to get the very best quality sawn timber in a host of different dimensions. One machine, called the gang saw, has thirty-two saws arranged in a vertical row and can quickly reduce big section timber into a great number of boards simultaneously. As can be imagined, the saws need constant attention, and high above the floor of the mill is the saw shop where an expert 'saw doctor' and his staff are at work.[1]

Brandler and Rylke work an average of four to six square miles of rain-forest a year in the reserves and about three square miles outside the reserves, where trees are purchased directly from villagers. The combined lumbering and sawmill operation employs some 700 workers. Apart from money received in wages, there are also direct benefits to the community in terms of roads and bridge building. The Obubra–Ediba road has been widened and tarred to aid the evacuation of timber, and the Oferekpe–Ogada ferry is used mainly by Brandler and Rylke trucks which pay a substantial toll to help keep it running.

At present sawn timber is stored in the company's yard in Enugu, whence shipments are made by rail to Port Harcourt for export or to internal markets. The less durable timber is treated against insect and fungal attack. An expansion programme includes the establishment of a veneer mill, also plywood and chipboard factories, but such developments will depend on a considerable improvement in transport facilities.[2] The proposed Ikom–Calabar road, with its associated linkage with the Enugu–

---

[1] African Timber and Plywood (Nigeria) Ltd., *A.T.P. Sapele* (n.d.), p. 5.
[2] Anon, 'Rural Industrialization – Timber Industry', *Eastern Nigeria*, vol. i, No. 3 (Dec. 1963), p. 26.

Abakaliki network, will provide the necessary improvements in communications sought by the timber firms.

(v) MANUFACTURE OF CEMENT AND ASBESTOS CEMENT PRODUCTS

In 1954 one of the most ambitious industrial ventures to be launched in Nigeria at that time was initiated at Nkalagu, a small village some thirty-two miles east of Enugu. Where there was once little but bush and patches of farmland now stands a modern manufacturing plant capable of working twenty-four hours a day, seven days a week, and employing nearly one thousand people on shift work. The Nigerian Cement Co. Ltd. has an annual output of some half a million tons of high-quality Portland cement, an essential commodity for the vigorous building programme in the Eastern Provinces. Due to the resounding success of the Nkalagu operation, import of cement from abroad has been drastically reduced over the last decade.

The two main requisites for the economic production of cement are abundant supplies of raw materials and fuel. At Nkalagu there are readily available good-quality deposits of limestone and shale. Only gypsum, of which a small amount is used in producing cement, has to be imported. As for the source of power, coal is supplied from the nearby Enugu collieries, while electricity is provided from the Oji River generating station, some sixty miles to the west. The factory is linked by an eight-mile railway spur from the main line, also by road to Enugu.

The processing plant for Nigercem comprises a number of imposing structures: storage bunkers, slurry silos, slurry basins, kilns, cement silos, hopper and packing facilities, and so forth. There is also machinery for manufacturing the many millions of paper bags required annually for shipping the cement. Adjacent to the industrial complex are extensive housing estates, consisting of over four hundred company-owned houses of various types and sizes for employees. Running through the township are seven miles of roads, providing access to residential quarters, clubs, schools, dispensary and hospital, and the police station.

Closer to Enugu, at Emene, an ultra-modern plant for the production of asbestos cement building materials was opened in 1963. Turner Asbestos Cement (Nigeria) Ltd. utilises locally-produced cement from Nkalagu, and imported asbestos, a mineral with unique qualities of fire and weather resistance and insulation. The comparatively large amounts of water required in the manufacturing operation are drawn from the nearby

Ekulu River. It is pumped to the factory where it is filtered and aerated, and its pH value corrected before use.

The installed machinery is the most modern and advanced of its kind in the world. Products include roofing sheets of various profiles, moulded items such as water cisterns, septic tanks, ventilator flue fittings and roofing accessories. Conscious of the growing need for materials with which to build better houses in villages and towns, the company has developed new products for low-cost housing including sink units, squat toilets and double-width lightweight corrugated sheets. Increasingly on the landscape, the light-grey, red- and green-tinted sheets of roofing material are vying with corrugated-iron and alluminium roofing sheets in replacing grass thatch and palm fronds on residences and business buildings. The most recent development is the production of asbestos cement pressure pipes and accessories with which the region-wide plan to provide a piped water supply to all communities may be accomplished.

(vi) IRON AND STEEL PRODUCTS

A significant step was taken in the industrialisation of Eastern Nigeria in 1962 when a small steel rolling mill was opened at Emene, seven miles east of Enugu. The mill utilises iron and steel scrap which is supplied from all parts of Nigeria and produces mild steel rods, high tensile steel bars, wire rods, angle iron, flat iron, channels and other sections used in building construction. The scrap metal is fed into an electric arc furnace which receives power from the Oji River and limestone flux from Nkalagu. The railway distributes the processed metal products throughout the Republic. Annual output in 1966 was in the order of 12,000 tons of rods and other items.

The Nigersteel Company Ltd., like other factory-type industrial establishments in Eastern Nigeria, is fulfilling several functions. It not only produces steel for constructional and other industrial uses, but it also provides employment and training in the difficult field of metallurgy. Crane drivers, mill hands and artisans are acquiring skills entirely different from those inherited from their peasant farming parents in the rural areas. Such skills will be of value in a wide range of future metal-working industries. Furthermore, the Nigerian staff trained in specialised fields in the steel industry will form the nucleus of indigenous staff of the proposed all-Nigerian integrated iron and steel plant.

The urge to establish a basic iron and steel industry within Nigeria arose soon after independence in 1960. Apart from its prestigous value the new leaders believed that not until the country had established a steel plant could real industrialisation be said to have begun, since iron and steel are fundamental requirements in nearly all other industries. 'Once this industry is started we shall shake off the inertia of centuries and advance like other modern nations.'[1]

The initial hope in Eastern Nigeria was that the iron and steel plant might be located at Enugu, on the railway and in close proximity to the coal, the Nsukka-Udi Plateau iron ore, and the Nkalagu limestone. The inferior qualities of the coal and ferruginised ironstone have already been commented upon. Inter-regional rivalries and political decision-making were to influence the choice of location, however. In May 1964 the National Economic Council announced that it had decided to establish two separate complexes, one in Northern Nigeria and one in the East. A blast furnace was to be located at Lokoja, in the proximity of abundant limestone and some 160 million tons of 45–50 per cent iron-content ore (the reserves of ore west of Enugu amount to approximately 60 million tons of 32 per cent iron-content). The second complex, a rolling mill, was to be located at Onitsha. There is no rail or easy road connection between the two sites, but both towns are on navigable sections of the Niger and presumably raw steel could move by water to Onitsha from Lokoja, and finished products could travel by water or road to consuming centres. The proposed yearly throughput was to be around 200,000 tons (the estimated annual consumption of iron and steel in Nigeria by 1970), with production concentrated on black and corrugated sheets, rods and bars, angles and channels, besides other structural steel items.

Between 1964 and the military *coup* in January 1966, no serious work on establishing the two plants was undertaken. An outline agreement between the Nigerian Government and a consortium of American, British and German firms for the establishment of an iron and steel industry was signed only in August 1965.[1] In May 1966, the Military Government announced their intention to proceed with the proposed iron and steel project, but by 1967 the Civil War had further postponed initiation of the scheme.

On strictly economic grounds, the wisdom of developing an indigenous iron and steel industry in Nigeria may be questioned. Steel plants are very expensive to erect. Modern steel plants have an optimum production of 1 million tons a year, far in excess of Nigeria's needs. Moreover, not all kinds of steel can be economically produced by a single plant, so the effective domestic market for a Nigerian iron and steel works is less than

---

[1] Speech in 1960 by Dr M. I. Okpara, former Premier of Eastern Nigeria.
[2] Anon, 'New Nigerian Giant?' *West Africa*, No. 2517, 28 Aug. 1965, p. 953.

200,000 tons a year. For an industry such as iron and steel which requires a large market, the most richly endowed country in West Africa in terms of resources might better be responsible for meeting the needs of the entire area through a suitably arranged customs union or 'common market'. Thus Liberia or Guinea, with their vast reserves of high quality iron ore, could well support an integrated steel plant to meet the needs of all West African countries. Unfortunately political realities tend to negate the economic factors of location in strategic issues of this order of importance.

Nevertheless, agreement to create a West African Iron and Steel Authority was reached in Monrovia in 1965, the signatories being Nigeria, Liberia, Ghana, Mali and Guinea. It is to be hoped that some measure of co-ordination and co-operation can be achieved in West African planning for heavy industry.

(vii) TYRE MANUFACTURING

From the end of the nineteenth century, when the first motor vehicle was introduced into Nigeria, the movement of people and products by road has grown at a rapid rate; the number of vehicles – cars, lorries, buses etc. – has increased in a particularly pronounced fashion since 1960, with the opening up of rural areas to trade and industry. As the economy continues to expand under development planning, with more industries established, agricultural production expanded and additional roads constructed, so the need for further wheeled vehicles will increase. This expansion in the road transport industry means a corresponding growth in the consumption of the products of the tyre manufacturing industry.

The recognition of this growing need for tyres, inner tubes and accessories, together with the fact that Eastern Nigeria is fast becoming an important producer of natural rubber, led to the decision of a leading European tyre manufacturer to establish a plant in Port Harcourt.

The first vehicle tyre made in Nigeria was produced in September 1962 at the factory of Michelin (Nigeria) Ltd., in the Trans-Amadi Industrial Estate.[1] Within a year, the first shipment of Nigerian tyres was made to Europe and since 1963 Michelin tyres have been exported to Europe, Asia and other parts of Africa. The factory itself is attractively designed and equipped with the most modern materials and machinery. The Michelin plant was the first to utilise natural gas from the Shell–BP oilfields for

[1] Anon, 'Tyre Industry in Nigeria', *Nigerian Trade Journal*, vol. xii, No. 2 (April/June 1964), p. 48.

process steam-raising in the factory. In addition to rubber from Nigerian plantations, cotton fabric is supplied by Nigerian textile factories, while synthetic rubber, nylon, bead wire, carbon black, sulphur and zinc oxide are imported from the U.S.A. and the U.K.

The industry employs over 500 people in the production, engineering and administrative divisions, and there is a valuable training programme for Nigerianisation of key staff. Many qualified Nigerians now hold executive positions in the tyre manufacturing industry. It also creates a steady market for Nigerian natural rubber and serves as a protective shield both against competition from Liberia, Malaya and other producer countries, and against the uncertainties of the world market.

16 The Main Urban Centres of Eastern Nigeria and their Industrial Importance

INTRODUCTION

THE geographical location of the most important large-scale industries in Eastern Nigeria is shown in Fig. 16.1. As indicated in this map, the six towns of Port Harcourt, Enugu, Onitsha, Aba, Calabar and Umuahia contain almost all the modern industrial establishments.

This concentration is in response to a number of locational advantages (raw materials, power supply, transport facilities, semi-skilled and skilled labour, markets, etc.) but is also due to systematic planning for industrial growth on the part of government. The Ministry of Town Planning and Surveys, for example, has established Planning Authorities for all the large towns, which encourage local communities to undertake land development schemes and act as intermediaries in providing the requisite technical advice. This advisory service extends particularly to industrialisation and the location of new industries.

Recently the Government has initiated a series of industrial estates in the major towns of the region, placing their development in the hands of the Eastern Nigeria Industrial Estates Co. Ltd. This company is charged with the responsibility of adminstering the layouts, facilitating the establishment of new factories, and granting concessions in terms of rent reductions on industrial premises and houses built by the Eastern Nigerian Housing Corporation.

A brief account of the principal urban centres now follows, describing their growth, locational advantages and handicaps for industry, and overall importance to the economic development of Eastern Nigeria.

PORT HARCOURT

Port Harcourt lies about forty-one miles from the sea up the Bonny River. It is a relatively new town, being a creation of the colonial period, and was first surveyed in 1912. It is generally believed to be named after a Colonial

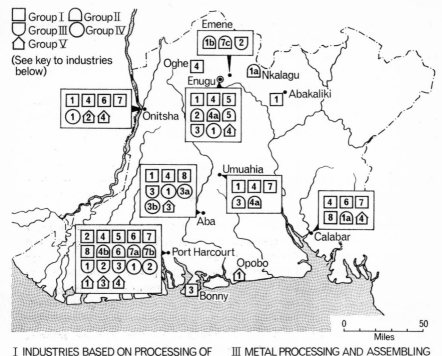

Key:

- Group I
- Group II
- Group III
- Group IV
- Group V

(See key to industries below)

I INDUSTRIES BASED ON PROCESSING OF AGRICULTURAL AND TREE PRODUCTS

1 Beverages
2 Cigarettes
3 Fibre processing: coir
4 Foodstuffs
5 Furniture
6 Rubber products
7 Saw milling
8 Vegetable oil processing

II EXTRACTIVE (Mineral-based) INDUSTRIES

1a Cement   b Asbestos cement products
2 Coal mining   3 Drilling clay
4a Ceramics   b Glass products
5 Lime (fertiliser)   6 Paints
7a Natural gas   b Petroleum   c Industrial gases

III METAL PROCESSING AND ASSEMBLING INDUSTRIES

1 Aluminium products
2 Iron and steel products
3 Lorry assembly, repair, auto-engineering

IV CONSUMER GOODS

1 Textiles and footwear
2 Enamel ware
3a Pharmaceuticals   b Cosmetics

V MISCELLANEOUS INDUSTRIES

1 Boat building   3 Plastics
2 Gramophone records   4 Printing

FIG. 16.1 *Location of selected large-scale industries (Eastern Nigerian Surveys, 1965)*

Secretary of the day, Sir William Harcourt, but another version has it that the name derives from a Marine Department navigator, Mr D. L. Harcourt, who was allocated land in the area in 1911.

Some years after the town was marked out, construction of the railway from Port Harcourt to what were then the Udi coalfields greatly enhanced the importance of the town, which rapidly became one of the principal centres of trade distribution and completely eclipsed the old indigenous port of Bonny at the mouth of the river, famous in earlier days for its domination of the slave and oil palm trades. Today, Port Harcourt is the second largest port in Nigeria, after Lagos, and is the major port for Eastern Nigeria; with a population of 180,000 in 1963, it is also the largest urban centre in the Eastern Provinces. By rail and road it serves as the outlet for the bulk of Eastern Nigeria's exports of agricultural and forest products, other raw materials and manufactured goods, also a substantial portion of Northern Nigeria's exports. Similarly the greater part of the imports into Eastern Nigeria and much of Northern Nigeria also passes through Port Harcourt.

Unfortunately the Bonny River is not navigable by large ocean-going vessels; only with the recent dredging of its bar to facilitate oil exports, and subsequent dredging of the main channel upriver, have 18,000-ton ships been able to reach Port Harcourt. The berthing facilities are also restricted and considerable congestion of shipping is experienced due to the general configuration of the harbour. A major port extension programme was completed in 1960, providing a total of eight berths and seven transit sheds, but this has already proved inadequate to cope with the growing volume of traffic and trade. Further extensions to quays and storage facilities are under way. The town itself is sited on low-lying land at the edge of the salt-water swamp of the eastern Niger Delta, broken up by numerous tidal inlets such as the Amadi and Dockyard Creeks. The limited amount of firm dry land close to the harbour has hampered commercial development and influenced the unusual morphology of the town.

Nevertheless down-town Port Harcourt is a scene of intense commercial activity. Many trading companies have offices along the roads near the harbour and railway terminus, particularly in Azikiwe Road, Station Road, Harbour Road, Industry Road (Table 16.1). High density residential areas also abound in the southern section of the town.

The main industrial zone – the Trans-Amadi Estate – is being developed some four miles north of the port area. This industrial layout covers an area of 2,500 acres; 1,500 acres are for industrial plots, ranging from 1·3 to 4·0 acres (without rail sidings) and from 2·2 to 33·0 acres (with abutting rail sidings); 500 acres are for a workers' residential area, when fully developed, to house over 20,000 people, while the remaining 500 acres are taken up by roads, parking areas and railway sidings.

The factory section is subdivided so that plants which create noise, smoke and odours are separated from lighter industries with no pollution problems. Living and manufacturing areas are also separated by belts of office buildings or landscaped grounds, while two miles west of the industrial estate a further low density residential area is being developed

TABLE 16.1

Selected list of important businesses located in
central Port Harcourt (1965)[1]

I. *Building and civil engineering contractors*
1. Costain (W.A.) Ltd.
2. Micheletti & D'Alberto Ltd.
3. Parkinson (Sir Lindsay) Nigeria Ltd.
4. Taylor Woodrow (Nigeria) Ltd.

II. *Electrical engineers and contractors*
5. Rashleigh Phipps & Co. (W.A.) Ltd.
6. The West African Engineering Co. (Nigeria) Ltd.

III. *Importers*
7. Allied Trading Co. Ltd.
8. Hansa Nigerian Trading & Motor Co. Ltd.
9. Sick-Hagemeyer (Nigeria) Ltd.
10. U.A.C. (Technical) Ltd.
11. Witt and Busch Ltd.

IV. *Merchants*
12. Chanrai & Co. Ltd.
13. Chellaram & Sons (Nigeria) Ltd.
14. Compagnie Française de l'Afrique Occidentale
15. Holt (John) Ltd.
16. Kingsway Stores Ltd.
17. Leventis & Co. (Nigeria) Ltd.
18. Ollivant (G.B.) (Nigeria) Ltd.
19. Société Commerciale de l'Ouest Africain
20. Union Trading Co. Ltd.

V. *Motor dealers*
21. Armels Transport Ltd.
22. Incar Motors Ltd.
23. Leventis Motors Ltd.
24. Mandilas and Karaberis Ltd.
25. Niger Motors Ltd.
26. Société Commerciale de l'Ouest Africain
27. Union Trading Co. Ltd.

VI. *Petroleum products distributors*
28. BP (West Africa) Ltd.
29. Esso West Africa Inc.
30. Mobil Oil Nigeria Ltd.
31. The Shell Co. of Nigeria Ltd.

VII. *Shipping agents*
32. African & Overseas Agencies Ltd.
33. Elder Dempster Agencies Ltd.
34. Holt (John) Shipping Services Ltd.
35. Intercotra Ltd.
36. Palm Line (Agencies) Ltd.
37. Scandinavian Shipping Agencies Ltd.

VIII. *Tobacco manufacturers*
38. Nigerian Tobacco Co. Ltd.

[1] Port Harcourt Chamber of Commerce, *Port Harcourt* (Port Harcourt, n.d.), pp. 14–29.

by the housing corporation around the luxury Presidential Hotel. With provision for a shopping centre, bus terminus, clinic, police and fire stations, the Trans-Amadi Layout will eventually emerge as a distinct urban centre in its own right.

The topography is excellent for industrial development; gradients are about two to three per cent and soil tests have proved satisfactory for the foundations of heavy machinery. Electrical energy comes from the 210,000 kW. station at Afam, generated by natural gas, while a two-mile pipeline from Apara conveys gas as an alternative power source. Chlorinated and potable water is supplied to the industrial area at the rate of $3\frac{1}{2}$ to 4 million gallons a day.

Important industries already established in the Trans-Amadi Layout include Alcan Aluminium of Nigeria Ltd., Apex (Eastern Nigeria) Ltd. – paper and stationery, Michelin (Nigeria) Ltd., Nigerian Glass Co. Ltd., Whessoe Engineering Ltd.; others are under construction.

ENUGU

The capital of Eastern Nigeria is a further illustration of a European-originated community founded soon after the British occupation. In 1909 a geological party seeking silver along the Udi Escarpment located instead the coal formations of the Lower Coal Measures. Five years later work was begun on building a railway to link the Udi coalfields with the coast at Port Harcourt. Enugu developed rapidly with the completion of the railway in 1916. The town takes its name from the village of Enugu-Ngwo which overlooks Enugu from the scarp on the western margin of the town. 'Enugu' is a contraction of the Igbo words 'enu Ugwu' which means the top of the hill, and Enugu-Ngwo is the hill village of the Ngwo people. There are several 'Enugus' in Iboland, but use of the place name today invariably signifies the administrative centre of Eastern Nigeria.

Enugu grew rapidly and in 1929 it became the headquarters of the Southern Provinces, taking over this political role from Lagos. When in 1939 the Southern Provinces were split to form the Eastern and Western Provinces, Enugu became the Eastern headquarters. After 1951, the Eastern Provinces were renamed the Eastern Region, and Enugu was developed further as the administrative seat, with buildings for the Regional Legislature (the House of Assembly and the House of Chiefs), the official residence of the Governor, the Regional Ministries, the Statutory Corporations and other governmental agencies.

Over the years, the political functions of Enugu remained paramount and, apart from coalmining and a railway maintenance workshop, little large-scale industrial development took place until after independence. Nevertheless the combination of abundant supplies of coal, usually adequate water and electricity services, rail and road facilities and a sizeable labour force make Enugu an attractive area for certain industries; up until 1967 considerable efforts were being made to attract new manufacturing plants to the capital. The chief disadvantage of Enugu's location is that it is somewhat off-centre so far as internal markets are concerned, being sited to the north of the main concentration of population in Owerri and Uyo Provinces.

The town is situated in the undulating, dissected piedmont section of the Udi Escarpment, about 745 feet above sea level. It comprises a number of discrete layouts with gridded street patterns, divided by incised valleys and streams. The older sections comprise Iva Valley, Ogbete, Ogui, Asata, Uwani and the Government Residential Area; they contain many small business establishments, shops and living quarters, the latter varying from small inferior houses in Ogbete and Asata to spacious old edifices in the G.R.A. With the rapid growth of population, Enugu has expanded through new residential districts, notably Independence Layout, with the expensive, imposing homes of former Cabinet Ministers around the Presidential Hotel, Ekulu Layout, Aria River Layout, Ogui Nike Layout, Uwani Northern and Southern Extensions. Many new houses and flats have been constructed in recent years to meet the needs of workers, civil servants and the managerial executive class. With a population of over 138,000 in 1963, Enugu is the third largest town in the Eastern Provinces (after Port Harcourt and Onitsha).

In a deliberate move to encourage industrialisation, an industrial estate has been laid out at Emene, some seven miles east of Enugu on the Abakaliki road (near the airport), and three important factories were already in operation by 1963: Turner Asbestos Cement (Nigeria) Ltd., Nigersteel Company Ltd., and Nigergas Ltd., which produces oxygen and acetylene for industrial purposes. The Nigerian Cement Company, operating at Nkalagu, thirty-two miles east of Enugu, may for the purpose of this survey be included with the industrial enterprises in the capital. It depends upon Enugu for coal for firing the kilns and supplies one of the two essential ingredients processed by the Turner Asbestos Cement Co. at Emene.

Within Enugu itself, the Eastern Nigeria Furniture and Construction Co. and the Ekulu Pottery are new enterprises located in the Iva Valley suburban section. The Ministry of Works runs a workshop and sawmill while a Government Trade Centre (G.T.C.) and Government Technical Institute (G.T.I.) provide apprenticeships and important technical education for industrialisation. Smaller enterprises include lorry-building and vehicle repairing, cabinet-making and joinery, shoemaking, gold-working, printing, metalworking, textiles, tailoring, etc. A pioneering estate for small indigenous industries was created in Uwani Northern Extension in 1965–6.

ONITSHA

The riverine town of Onitsha is believed to have been founded at the beginning of the sixteenth century by a group of emigrants from Benin who crossed the Niger and settled on its eastern bank. It was not for another three hundred years that the name 'Onitsha' acquired more than local significance when, as a result of European penetration and commercial expansion upriver, traders began to establish stations on the banks of the Niger. In 1863 trade treaties between the British Consul for the Bight of Benin and Biafra and the Obi and Chiefs of Onitsha enabled trading companies to establish themselves on a firm footing in the town.

By 1966 Onitsha was a thriving community with a population of over 160,000 (the second largest urban centre in Eastern Nigeria), and recognised as one of the most important indigenous trading centres in the whole country.

> Onitsha has been described as the 'realm of the merchant princesses', a title which acknowledges not only the business acumen of its women traders but also the vast scale of their operations and the importance of the town as a distribution centre. The finest and most modern market in Nigeria is at Onitsha, costing over £500,000 and containing more than 3000 stalls.[1]

The town acts as an entrepôt port between the Delta ports along the coast and other trading stations on the upper Niger and Benue rivers in the interior. Year-round navigation of the Niger River is possible only as far as Onitsha. This has helped to foster the growth of the town as a trans-shipment centre. Powered river craft based at Warri and Burutu in the mid-west ply between the Delta ports and Onitsha, thence to Lokoja and Makurdi and, for a very limited period, as far as Baro on the Niger, and Yola and Garua on the Benue. Onitsha comes into its own as an entrepôt port during the low-water season from January to June, when bulk cargoes from the Delta ports and produce from the Northern Provinces are trans-shipped at Onitsha.

[1] Nigeria, Federal Ministry of Commerce and Industry, *This is Nigeria: Onitsha* (Lagos, n.d.). The market was destroyed during hostilities in 1968.

The town is also the gateway by road from Lagos, the Western and Mid-Western Provinces to the Eastern Provinces of Nigeria, and a centre from which goods are distributed via a dense network of roads throughout the East and as far as the Cameroon Republic. The chief locational disadvantages are that there is no airport or rail service to Onitsha and the nearest facilities of these transport media are at Enugu, sixty-seven miles to the east. Road communications were also severely handicapped until recently by the absence of a bridge over the Niger. The Inland Waterways Department operated a ferry service with eleven double crossings a day but the service was notorious for its congestion and the associated costly delays experienced by waiting lorries and cars.[1] Fortunately this situation was remedied by the completion in December 1965 of a superb bridge across the Niger River south of Onitsha (damaged by the war in 1968). The street pattern of central Onitsha, unlike that of Enugu and Port Harcourt, is unplanned and reflects its pre-colonial origins. Only south of Otumoye Creek is there a clear indication of urban planning in Nupe Square and associated streets.

Traditional industries in the Onitsha area include sawmilling, cabinet-making and joinery, shoemaking, tailoring, bus-body building, motor repairing, printing and baking. Newer enterprises involve tyre retreading, soft drinks, the manufacture of gramophone records, also plastic travel bags and accessories. The bridgehead site is being developed as an industrial estate and a large textile plant owned by Textile Printers (Nigeria) Ltd. commenced production in 1966. The proposed iron and steel rolling mill is also to be located in the vicinity of the Niger bridge.

ABA

The township of Aba is a creation of the twentieth century, and emerged on a site formerly occupied by several village communities. In 1901 a military camp was established by the British to check civil disturbances, and to serve as a base for the expeditionary force which was to invade Arochuku and destroy the 'Long Juju' in the following year.

The Aba area was of some agricultural and commercial importance even before the Europeans arrived. Its central position in the north-west to south-east axis of high population and proximity to the Aba River

---

[1] In September 1964 the author and his family in a private car waited in line for over fourteen hours before obtaining a place on the ferry. Lorries not infrequently were delayed several days while waiting to cross the Niger.

attracted traders from many other communities, particularly Arochuku, Awka, Nnewi, Bende, Nkwerre and the coastal town of Opobo. Following the military occupation, Aba's significance as a trading centre was greatly enhanced. Construction of the thirty-six-mile rail link with Port Harcourt in 1915 and extension of the railway to Enugu in 1916 assured Aba of its dominance over other traditional commercial centres. By the 1930s Aba was firmly established as an urban area, with a growing diversity of trading and small manufacturing interests.

Today Aba is the fourth largest town in Eastern Nigeria, with a population of over 130,000. It remains one of the most important commercial centres in the region and continues to attract new industries due to good rail and road facilities, permitting entry of raw materials and evacuation of finished products, adequate electricity and water supplies, an abundance of flat level land, good supply of unskilled and semi-skilled labour, and a large local market. It enjoys in many respects the locational assets of Port Harcourt yet avoids the restrictions of a low-lying, swamp-surrounded site.

Aba is a compact and well-planned community. A large area for industrial development has been set aside between the Aba River and the railway sidings to the north of the town and several important industries are already located in the estate, including Star Nigerian Breweries Ltd., Aba Textile Mills Ltd., Pfizer Products Ltd. (pharmaceuticals), two soap factories, a cosmetics factory, a bakery, furniture workshop, a plastics factory, and a technical school for motor engineers. Elsewhere in the town there are textile and clothing plants, a raffia industry, cabinet-makers and joiners, printers, metal workshops and many other small industries.

OTHER TOWNS OF GROWING INDUSTRIAL IMPORTANCE

CALABAR

Calabar (76,410) is among the oldest and most historic of Eastern Nigerian towns. In its heyday it was a prosperous slave-trading port and later, from 1883 to 1906, became the headquarters of the Oil Rivers Protectorate. During the present century, however, its importance has progressively waned with the rising commercial importance of Port Harcourt and the administrative strength of Enugu. After 1960 the Efik town failed for political reasons to receive governmental support in the development of new industries. The eccentric location of Calabar in relation to the other centres of urban and rural population in Eastern Nigeria, as well as loss of the hinterland territory of the British Cameroons, also contributed to the atrophying of Calabar's historic trading and political importance.

Nevertheless the future for Calabar appears much brighter for a number of reasons. The need to utilise the lightly-settled areas and underdeveloped resources of Calabar Province and the Upper Cross River Basin came to be recognised by the former civil government and is no doubt appreciated by the present military authorities. The agricultural potential is already being realised through the concentration of plantations in the immediate hinterland of Calabar. Timber and mineral resources have hardly been tapped as yet. The most significant development is however the proposed Calabar–Ikom road (Chapter 17) which will benefit the fortunes of the town immeasurably and help to restore some of Calabar's former influence.

Located some forty-five miles up the Cross and Calabar Rivers, Calabar is about the same distance inland from the open sea as Port Harcourt. It is Eastern Nigeria's best natural deepwater port and further dredging of the Calabar River could bring to Calabar vessels of greater tonnage than those reaching the premier city of Port Harcourt. An indication of Calabar's renaissance is the imposing structure of a new £3,500,000 cement factory, supervised by Fritz Werner *G.m.b.H.*, a West German industrial group. Feasibility studies involving new industries for producing plywood, ceramics and starch have been undertaken. Expansion of Calabar's traditional commercial and industrial activities – boat-building, rubber-processing, and the export of palm produce, cocoa, rubber and timber – may confidently be expected in the future.

UMUAHIA

This is another European-promoted town on the railway between Port Harcourt and Enugu, with a favourable location in the centre of the region. While supporting the usual range of small indigenous industries, and a novel all-Nigerian enterprise, the manufacture of drilling clays and paints, Umuahia could boast of no large-scale, factory-type enterprises until 1963 when the ENDC-backed Independence Brewery was established in the town. With the Nigerian Brewery plant just thirty-nine miles south in Aba, this duplication of facilities is somewhat uneconomic but the Golden Guinea brewery in Umuahia is a Nigerian as distinct from a foreign-controlled enterprise and its location in Umuahia is a clear illustration of political decision-making in industrial planning. A second modern manufacturing establishment in Umuahia is the £500,000 Modern Ceramics Industries Ltd., also financed by the ENDC, and producing dinner ware and sanitary fittings.

OWERRI

This town is an old government station whose political authority and economic growth have been dwarfed by Port Harcourt, Aba and Umuahia, all previously in Owerri Province. Port Harcourt, and later Umuahia Province, were carved out of Owerri Province, and the two towns withdrew much of Owerri's administrative functions by becoming Provincial capitals in their own right.

Owerri is off-centre so far as the main north–south line of road and rail communications between Port Harcourt and Enugu are concerned, and has failed to benefit from economic developments along this transport axis. Hopes of a revival of importance were raised when Shell–BP Petroleum Co. Ltd. established its headquarters at Owerri in the 1950s, but the subsequent transference of this strategic company to Port Harcourt effectively dimmed the vision of an industrial complex in the town. The location of an Industrial Development Centre for improving the skills of local entrepreneurs and craftsmen in woodwork, leatherwork and metalware, and motor mechanics; an Advanced Teachers' Training College on the former Shell–BP site; a new ENDC 'Progress Hotel'; and substantial road improvements may encourage the expansion of this otherwise modest town.

17 The Pattern of Communications in Eastern Nigeria

INTRODUCTION

THE canoe, motor vessel, railway engine and lorry have successively and together played a major role in the economic development of Eastern Nigeria. Each still has their part to play although it is the road transport system which is expanding most rapidly and has become the most significant component of the 'geography of circulation' in the Eastern Provinces. The newest addition to the communications network is the Nigerian Airways Corporation, although a fuller contribution to the movement of both people and freight by air transport still lies some distance in the future.

ROADS

In 1900 virtually no motorable roads existed in Eastern Nigeria with the exception of a small network in the Calabar area. Very little road building took place during the early years of the British occupation. With the construction of the railway during the First World War, care was taken to see that roads should not compete with the line of rail; most of the roads which were built served as feeders to the railway system. By 1920 it was possible to motor from Port Harcourt to Onitsha, thence to Enugu, with a perilous descent of the escarpment overlooking the coal-mining community to the west. By 1927 a trunk-road policy had been formulated, and roads ceased to be constructed mainly as feeder supply lines to the railway. Competition between the road and rail systems for freight henceforth became keen and continues so to the present day.

This policy, which has been amended only slightly since that date, classifies the roads under three groups:[1]

(i) Trunk roads 'A', which are constructed and maintained out of central Government or Federal funds. The basic aim of the 'A' roads is to

[1] Anon, 'Road Construction', *Nigerian Trade Journal*, Special Independence Issue, 1960 (Lagos, 1960), p. 52.

connect Lagos with the other administrative centres of the country, and to link these towns with each other and with the ports.

(ii) Trunk roads 'B', which connect provincial or district headquarters and other large towns with the trunk 'A' system, with one another, and with ports or points on the railway. These roads are constructed out of Provincial or Regional funds and maintained by local government bodies with the assistance of grants from the Provincial government.

(iii) Most of the other roads, which carry mainly local traffic and act as feeders to the trunk road system; these are both constructed and maintained by local authorities.

Road building slumped in the early 1930s due to the economic setback, or 'Great Depression' in the Western world but was resumed on a substantial scale around 1937. During the Second World War road building throughout Nigeria was undertaken only to suit military requirements. After 1945 Colonial Development and Welfare Fund monies were made available for new road building and the Eastern Nigerian road grid began to assume its present pattern. In addition, a surge of community development projects enabled many miles of local roads to be constructed by voluntary labour.

With regional self-government in 1957, the new leaders were swift to recognise the value of a well-developed transport system to a growing economy. Transport was the largest component of the 1958–62 Regional Development Programme, taking up 28 per cent of the capital outlay of £20,732,400. In the four years covered by the programme, road mileage figures for Eastern Nigeria rose from 10,257 miles to 17,722 miles, an increase of 73 per cent (Table 17.1).

TABLE 17.1

Mileage of roads in Eastern Nigeria, by class of road and surface[1]

| Year | Bituminous surface | | | | Gravel or earth surface | | | | Total |
	Trunk A	Trunk B	Other	Total	Trunk A	Trunk B	Other	Total	(All roads)
1958	338	276	54	668	350	1,165	8,074	9,589	10,257
1960	495	500	63	1,058	193	977	11,802	12,968	14,026
1962	538	835	70	1,443	150	755	15,374	16,279	17,722

[1] Eastern Nigeria, Ministry of Economic Planning, *Statistical Digest of Eastern Nigeria* (Enugu, 1965), Table LIII, p. 54.

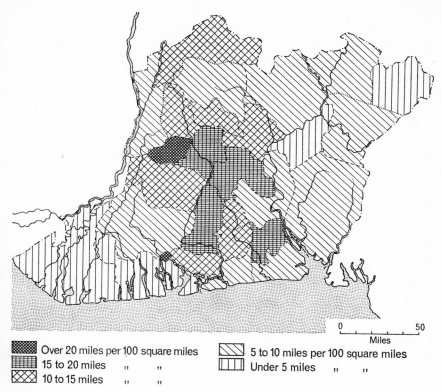

Over 20 miles per 100 square miles   5 to 10 miles per 100 square miles
15 to 20 miles   „   „   Under 5 miles   „   „
10 to 15 miles   „   „

0   50
Miles

FIG. 17.1   *Trunk road mileage per 100 square miles – by division, 1962*

The resultant density of roads (mileage of roads per hundred square miles of territory) in 1962 is shown in Fig. 17.1. As is to be expected, the revealed pattern closely reflects the distribution of population in the region as shown in Fig. 2.2.

In the Six-Year Development Plan 1962–8, transport was allocated some 11·8 per cent of the proposed expenditure of over £75 million: the third highest allocation after primary production, and trade and industry. This percentage was revised to 13·8 per cent in 1964 to pick up the costs of uncompleted road- and bridge-building projects from the earlier plan.

While the total road mileage will continue to grow it will not increase in so spectacular a fashion as during the period 1958–62. Much of the investment is going into improvements to the existing road network, especially realignment and resurfacing. Many of the earlier roads did not receive the benefit of engineering advice, were constructed of inferior materials, or built to meet low traffic densities. The marked increase in the number and weight of commercial vehicles has led to serious deterioration of the road surface and damage to narrow wooden bridges. The annual registration of vehicles in Eastern Nigeria increased from 3,640 in 1958 to 6,170 in 1962. Almost 45 per cent of the vehicles on the road in 1962 were lorries, 73 per cent of which were in the five-ton class or over, some even weighing twenty tons. Such heavily-laden lorries, fully loaded with palm fruits, timber or cement, rapidly take their toll of poorly designed and constructed highways. Despite such colourful and appealing slogans as: 'Remember Thy Promise O ! Lord'; 'God is My Leader'; 'Thy Will Be Done'; 'God Willing'; 'Be With Us O ! Lord'; 'Heaven Helps'; 'God is in Charge'; 'God's Case No Appeal'; 'Man of Peace'; and 'No Condition is Permanent', the accident rate due to inadequate road engineering and poorly maintained vehicles is alarmingly high. Present road development is therefore aimed at modernising the existing road transport system, widening culverts and bridges as well as broadening and straightening sharp bends.

The pattern of roads in Eastern Nigeria in 1966 is shown in Fig. 17.2. It is clear from the map that the Eastern Provinces have the best-developed road system in Nigeria and probably the densest network of any area of comparable size on the African continent. There is one mile of road for every 1·8 square miles of territory and, except for the riverine areas of the Niger Delta and the areas east of the Cross River, virtually every village in the region can now be reached by motor vehicles during the dry season.

While priority is being given to upgrading existing roads, new strategic roads are also being planned to open up hitherto undeveloped areas and resources or to improve on the old system. Two such roads are indicated in Fig. 17.2. The biggest of the new road projects is the Calabar–Ikom road, which is being financed by a loan of more than £4,500,000 from the United States Agency for International Development (USAID). This is the biggest single road project being supported by AID in Nigeria. The project includes a bridge over the Upper Cross River – from Ekurri to Adadama – which will provide a direct link between Enugu and Calabar.

The 'trans Cross River' area, which contains the bulk of the region's high forests, and is also one of the richest agriculturally, but is poorly served by motorable roads, will thereby be opened up for large-scale intensive development. The port of Calabar, nearest export route for

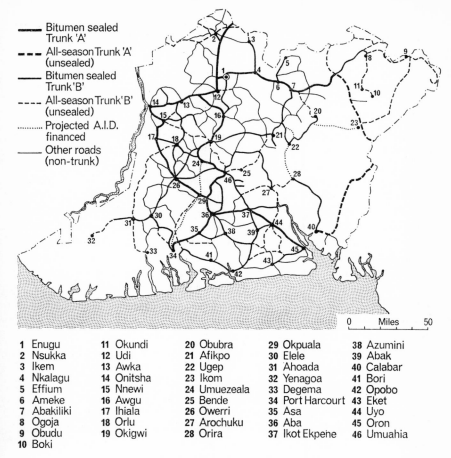

1	Enugu	11	Okundi	20	Obubra	29	Okpuala	38	Azumini
2	Nsukka	12	Udi	21	Afikpo	30	Elele	39	Abak
3	Ikem	13	Awka	22	Ugep	31	Ahoada	40	Calabar
4	Nkalagu	14	Onitsha	23	Ikom	32	Yenagoa	41	Bori
5	Effium	15	Nnewi	24	Umuezeala	33	Degema	42	Opobo
6	Ameke	16	Awgu	25	Bende	34	Port Harcourt	43	Eket
7	Abakiliki	17	Ihiala	26	Owerri	35	Asa	44	Uyo
8	Ogoja	18	Orlu	27	Arochuku	36	Aba	45	Oron
9	Obudu	19	Okigwi	28	Orira	37	Ikot Ekpene	46	Umuahia
10	Boki								

FIG. 17.2   *Pattern of roads, 1966*

the area, will also be served for the first time by a through road to its hinterland when the project is completed.[1]

Construction of the thirty-foot highway from Ikom via Ugep to the Calabar–Arochuku road was under way in 1966 and was expected to be completed by the end of 1968.[2]

USAID is also helping to finance another major road project north-

[1] Anon, 'Toward Better Services', *Eastern Nigeria*, vii (Sept. 1965), p. 2.
[2] Anon, 'The Missing Link', *Eastern Nigeria*, vii (Sept. 1965), pp. 16–19.

wards from Port Harcourt to Umuezeala (mid-way between Okigwi and Umuahia), estimated to cost £2,700,000. This road is planned as the southern portion of a central 'express-way' from Port Harcourt to Enugu and eventually Northern Nigeria, and is being designed as a multi-lane highway with a speed limit of 70 m.p.h. When completed, it will serve as a main artery for Eastern Nigeria's through traffic and, with connections to Onitsha – thence via the Niger Bridge to the Mid-Western and Western Provinces – it will expand existing markets as well as open up new trading areas both within and outside Eastern Nigeria.

The 4,600-foot-long Niger Bridge, sponsored by the former Federal Government and built by French contractors (Dumez) at a cost of £6,500,000, is a vitally important contribution to improved traffic flow between Eastern Nigeria and other parts of the country. It permits the rapid transit of lorries and passenger vehicles over a natural obstacle which through the years has greatly handicapped road communications in an east–west direction.

Also significant is the Mbiama–Yenagoa road in the Niger Delta, which was due for completion in 1966–7. It provides an all-season land route to the heart of the Delta, permitting a greatly increased commercial exchange between Ijaw communities and mainland towns which hitherto have achieved contact only by water.

THE RAILWAY

Except for limited stretches of dirt roads in Calabar and other townships, railroad construction preceded road building in Eastern Nigeria and played an early part in the progress and economic development of the region.

Despite the pioneering logistical and physical difficulties of construction through low-lying and ill-drained coastal plains, undulating hill country, high forest, palm bush and innumerable streams, the eastern rail link from Port Harcourt to Enugu was built with commendable, if not remarkable, alacrity between 1915 and 1916. Commencing from the embryonic settlement of Port Harcourt in 1915, the first train reached Aba over thirty-nine miles of track on 1 January 1916. A further sixty-seven miles of road upcountry to Afikpo was completed by 1 May, while the remaining forty-five miles to the coal-mining community of Enugu – the economic goal of this swift penetration of the so-called 'civilising rails' – were finished by 1 September 1916. A short branch line was constructed from the Enugu railhead (or Udi Junction as it was initially called) to serve the coalmines.

The rapid construction of the Eastern Nigerian railway is an excellent West African example of the comparable thrust into the interior elsewhere in the continent, e.g. Southern Africa, where mineral wealth awaited exploitation by European colonists. The continuation of the rail line northwards to tap the agricultural and mineral resources of Northern Nigeria was deferred until after the First World War. Work began in 1924, but Kafanchan and Jos were not reached until 1927 when the link-up with the Northern and Western sections of the railway enabled trains to make the lengthy and time-consuming journey from Lagos to Port Harcourt via Kaduna Junction.

Since completion in 1916 of the single track, 3′ 6″ gauge track traversing Eastern Nigeria from south to north, there have been no major extensions to this 'backbone' of rails in the region. Such additions as have occurred have been modest in mileage. Erection of the Nigercem factory at Nkalagu resulted in a nine-mile spur line to the factory from Ogbaho Station. Development of the Trans-Amadi Estate at Port Harcourt has resulted in a spur to meet the needs of this future industrial complex. A rail extension to the oil refinery at Alsea Eleme has facilitated the distribution of petroleum products overland within Nigeria. The prospect of a major new line to connect Port Harcourt and Onitsha, thence via the Niger Bridge to an east–west rail system in Western and Mid-Western Nigeria, has been mooted for some years. With the possible establishment of a steel rolling mill at Onitsha, the proponents of such a rail connection between East and West will no doubt intensify their efforts to achieve fruition of their objectives.

Such an ambitious building programme is quite unlikely to materialise, however (no provision was made in the Development Plan for such a project). The growing and increasingly effective road transport industry offers the most economic prospect for handling the expanding internal trade. The Niger Bridge incorporates no features to permit crossing of the river by rail. Development and extension of the recently completed trunk 'A' road from Benin to Lagos via Ijebu-Ode must take priority above any future rail link. The features of the terrain, with north–south orientated spurs of higher ground divided by many valleys would make construction of a railway in the west a prohibitively costly venture.

The only major extension to the national railroad grid undertaken in recent years is that of the 400-mile line from Kuru on the Jos Plateau via Bauchi and Gombe to Maiduguri, the so-called Bornu Extension. Completed in 1965, it is hoped that the new railway will stimulate agricultural development in north-eastern Nigeria, and much of the produce may eventually be expected to move via Enugu and Port Harcourt for export.

Despite competition from commercial road vehicles the train remains an important mode of transporting agricultural and industrial raw materials, manufactured products and people. The line of rail has been responsible for the growth of urban areas and the introduction of new industries along its path. It has aided the mobility of labour, particularly of young men in search of employment in the rail-serviced towns. An indication of the role of the Nigerian Railway Corporation in the pattern of communications in the country as a whole is given in Table 17.2, while Table 17.3 indicates the tonnage and relative importance of commodities carried from Eastern Nigerian stations in 1961–2.

TABLE 17.2

Nigerian Railway Corporation – passenger and goods traffic[1]

| | | Passenger Traffic | | | Goods Traffic | |
Period year ending 31 March	No. of passengers carried (thousands)	Passenger miles (thousands)	Average length (miles)	Tonnage hauled (thousands)	Net ton miles (thousands)	Average length (miles)
1953	5,516	351,213	63·7	2,086	827,246	397
1960	7,881	358,048	45·4	2,803	1,249,840	446
1963	12,006	515,905	42·9	2,760	1,410,950	511

[1] Nigeria Federal Office of Statistics, *Annual Abstract of Statistics Nigeria 1964* (Lagos, 1965), Table 6·4, p. 51. (Hereafter referred to as *Abstract of Statistics 1964*.)

TABLE 17.3

Tonnages of principal commodities carried from
Eastern Nigerian railway stations, 1961–2

TOTAL: 980,428 tons (100%)

I. *Mine and quarry products (tons)*			III. *Manufactured products (tons)*	
Coal	346,504		Cement	127,572
Salt	16,534		Petrol	21,953
Gypsum	15,013		Soap	9,710
Tin concentrate	8,572		Beer, spirits	9,067
Stone	4,954		Iron and steel	7,911
Columbite ore	2,015		Hardware	7,647
			Building materials	5,871
	393,592	(40%)	Vehicles	1,631
				191,362 (19%)

II. Agricultural products (tons)		IV. Forest products (tons)	
Groundnuts	73,934	Timber	27,295
Cattle	28,332	Pit props	10,230
Yams	24,251		
Cotton seed	21,605		37,525 (4%)
Garri	19,296		
Palm produce	13,793		
Cotton (ginned)	12,580		
Benniseed	11,328		
Oranges	6,602		
Guinea corn	6,069		
Sugar	6,050		
Bananas and plantains	5,540		
Flour	3,021		
Groundnut oil	1,985		
Rice	1,808		
	236,194 (24%)		

A recent and significant development on the Nigerian railways has been the introduction of diesel-electric traction. Diesel power is considerably more efficient and economical than steam-power and diesel-electric locomotives have brought substantial improvements in transit times. They were first introduced to Lagos in January 1955 and their use has since been consistent in Western and Northern Nigeria. In 1963, over a fifth of the mainline locomotives in use were diesel.

Their growing popularity has meant a sizeable reduction in the demand for Enugu coal on the railway system and this is one of the chief reasons for the difficulty in the coal industry. To try to maintain the demand for coal in the Eastern Provinces, all locomotives operating between Port Harcourt, Enugu and Makurdi were steam engines up until 1966. But the trend towards dieselisation is irreversible. International Bank experts have recommended that 'for more efficient and economic operation, steam traction should gradually be replaced by diesel-electric traction. . . . Basing our calculations on the experiences of other African countries, we estimate that aggregate savings of nearly 50 per cent can be realised on expenditure for fuel, water, engine operations and maintenance.'[1]

In the competition between road and rail, the Railway Corporation stands at a disadvantage. Road transport has many advantages over that of rail. The railway is a 'common carrier' which means that loads of all types must be accepted; the majority of road hauliers are not common carriers and can choose the most profitable loads while declining others.

[1] Anon, 'Railway Development', *Nigerian Trade Journal*, Special Independence Issue, 1960 (Lagos, 1960), p. 55.

F.E.N.

The Railway Corporation is obliged to publish its freight charges so that road transport competitors can inform themselves of these rates and adjust their own schedules accordingly to attract business.

Road transport enjoys the advantage of greater speed over short distances. The all-day journey from Port Harcourt to Enugu by train can be accomplished in six hours by lorry, five hours or less by Peugeot taxis. The use of road transport offers greater flexibility. Vehicles can go directly into the factory warehouse for loading and can take consignments straight to their unloading points. Rail shipment invariably requires delivery to the station by road, unloading and reloading into wagons, and the reverse procedure at the point of destination. Road vehicles can be used to take goods as soon as they are ready for delivery, whereas consignments by rail have to wait until a train is scheduled according to a timetable.

Technically, the lorry is easier to operate than a locomotive, requiring no 'firing up' as is the case with steam engines. Lorries are not restricted to a fixed track but can travel in any direction, wherever a negotiable road exists. There are, however, certain advantages enjoyed by railways over road transportation. Rail transport is quicker and cheaper for long-distance traffic, particularly if goods are sent by passenger train. Heavy lorries are slower on long-distance hauls; railway freight rates decrease in cost per mile as the distance increases, and this is of great advantage to the rail users.

Road transport has a limited (if often illegally exceeded) load capacity and cannot handle bulky materials as cheaply as the railway. For example, bulk shipment of coal, cement and petroleum in Eastern Nigeria is most easily made by rail, where a goods train manned by an engine crew of two and a guard can haul a large number of laden wagons simultaneously. Comparable shipments by road would require very many lorries and drivers. Special movements of machinery, heavy equipment and outsized materials for fabrication must perforce be made by rail. Imported processing plants for factories along the line of rail have all depended on the railway for their installation. The brewery vats at Aba and Umuahia, the asbestos-cement mixers at Emene and the cement kilns at Nkalagu are all illustrative of the indispensible functions of the Nigerian Railway. Rarely are trains delayed by heavy rains or flooding in the wet season, whereas many roads become impassable for lorries at the height of the rains.

Rail and road transport together, working in co-operation and with improved efficiency and organisation, will be responsible for sustaining the increased circulation of products from farm, forest and factory which economic development is bringing. Failure to cope with this geographic exchange of goods, services and personnel will seriously retard the goal of improved living standards for all Eastern Nigerians.

SHIPPING

INLAND WATERWAYS

Eastern Nigeria has extensive stretches of waterways in the creeks of the Niger Delta and the rivers draining the hinterland, of which the most important for navigation – apart from the Niger – are the Cross and Anambra Rivers. Inland shipping by small river-craft and canoes is an important economic feature of Eastern Nigeria's 'geography of circulation', connecting as it does communities which are inaccessible by rail or by main roads.

> Although history is somewhat obscure on the uses to which the waterways of the southern part of the country were put before the 17th century, records left by British explorers show that the value of the Rivers Niger and Benue as a means of penetrating the interior of the country and for the development of trade was quickly recognized. By 1857 river transport was gradually establishing itself and small paddle steamers of very shallow draft, not more than three feet, were operating in the Delta area and sailing far up the two rivers with goods.[1]

At the beginning of the twentieth century a number of creeks in the south were opened up and, to some extent, rivers charted; occasional dredging was undertaken and eventually a government department was set up to assume responsibility for the maintenance of such creeks and rivers as gave access to the trade and administrative centres.

With the construction of roads and an increase in vehicular traffic, it was necessary to establish ferries at points where the roads crossed rivers without a bridge, e.g. at Onitsha and Oron, where government- and privately-sponsored powered craft respectively carried lorries and cars to Asaba and Calabar. Away from the trunk roads, the ferry service consisted of rafts or pontoons poled by local inhabitants, as may still be found today at places like Itu, where the Enyong Creek remains to be bridged.

Over the years, small river ports developed and a clearer pattern of river trading emerged. Indigenous trading, consisting mainly of local and

[1] Anon, 'Development of Inland Waterways', *Nigerian Trade Journal*, Special Independence Issue, 1960 (Lagos, 1960), p. 49.

traditional produce such as yam, fruit, fish, cloth, pots, was still undertaken by the riverain people in their dugout canoes. Commercial produce for export as well as imported goods came to be the monopoly of slowly expanding river fleets, organised by expatriate interests and operating from bases established on the lower Niger and Delta. After the First World War the size of steamers brought into service increased and annual tonnages carried on the Niger and Benue increased appreciably. However, the success of river transport depended on good river seasons and reliable agricultural yields; unfortunately, these were by no means assured.

> The unpredictable rise and fall of the rivers coupled with the fact that produce was not available for down-river shipment until the end of the rainy season, in effect, at the end of the high-river season, adversely influenced the freight charges. River transport was thus faced with competition from both road and rail in view of the un-favourable conditions on the rivers, particularly the fact that fleets only operated for 30 per cent to 35 per cent of the year. Importers and exporters were discouraged from investing further capital on development with the result that these operations became greatly over-shadowed by road and rail developments and gradually fell into disuse.[1]

This was the picture until after the Second World War. With the flush of developments after 1945 the Nigerian trader – quick to recognise the beginnings of a new era – began to modernise his vessels by adding out-board motors to the dugout canoes or to replace them with shallow-draft motor barges, passenger launches and cattle ferries. These craft were cheaply constructed in government-sponsored boatyards and their purchase aided through loans. More frequent and regular services on creeks and rivers resulted, and a swifter exchange of goods and foodstuffs between main river towns and villages.

The larger river transport companies, encouraged by the economic trend and government support, also expanded their activities, investing a good deal of capital in their fleets and port facilities. Emulating developments in vessel design and operating techniques on other important rivers, e.g. the Rhine and the Mississippi, new and larger craft were introduced on the Niger. By 1958 it was not uncommon to see 'push-tow' trains of barges over 600 feet in length and carrying payloads of 3,500 tons at Onitsha.

The Nigerian Government also instituted studies into the nature and

---

[1] Anon, 'Development of Inland Waterways', p. 50.

behaviour of the main river systems. Teams of experts carried out hydrologic investigations of the routes with sea access, the main rivers and creeks, also the feasibility of constructing a multi-purpose dam on the upper Niger.[1] The Kainji Dam now under construction will, when completed, enable the main discharge of the Niger to be regulated, thus providing sufficient depth of water throughout the year for navigation at an economical draft. Ultimately the Niger and Benue will be opened up for continuous navigation from the sea to Nigeria's borders and beyond.

The Inland Waterways Department has also made great progress in recent years in improving navigational conditions for inland shipping.

> New buoyage systems have been introduced on the main rivers and special patrols, using up-to-date equipment for recording and broadcasting river depths and buoyage details, are operating on the Niger and Benue. River maps compiled from hydrographic surveys and aerial photography have been published, together with a network of river gauges, for the benefit of shipping. The success of these aids to navigation is reflected in the fact that the rapidly expanding river fleets now operate more effectively, not only during the high river season but at times of the year when formerly navigation was not possible.[2]

The Inland Waterways Department is thus contributing its own share to the economic development of Eastern Nigeria and the country as a whole.

SEA TRANSPORT

Nigeria has had some form of trade with the outside world for several centuries, but the end of the slave trade in 1840 saw the advent of legitimate maritime trade in agricultural products and manufactured goods. Steamships were introduced on the Europe–West Africa run in 1852 and in 1868 the British and African Steam Navigation Company was formed to organise a regular service. This company was the forerunner of the present Elder Dempster Lines Ltd.,[3] which has played a major part in developing sea communications between Nigeria and the Western world.

As the demand for Nigerian produce increased abroad, and as manufactured wares found a ready market within the country, so ships from other maritime nations – France, Norway, U.S.A., Japan, and more

[1] NEDECO, *River Studies.*
[2] Anon, 'Development of Inland Waterways', p. 51.
[3] Anon, 'Sea Transport', *Nigerian Trade Journal*, Special Independence Issue, 1960 (Lagos, 1960), p. 45.

recently Israel, India and Ghana – entered upon the scene. With this growth in shipping came a fuller realisation that the economic growth of the country was vitally dependent upon the efficiency of its ports, through which Nigerian produce was passing to the world markets, and where commodities essential for the existence of the country were being received for distribution by road, rail, and inland waterways. To improve the port facilities, the Nigerian Ports Authority was established in 1955 to organise harbour management and administration. The turn-around of ships was appreciably speeded up following the introduction of the N.P.A., although serious delays and congestion have continued to plague Nigerian ports with the rapid post-independence expansion of the country's economy.

Until 1959 the export and import trade of Nigeria was carried in ships owned and operated by foreign firms. In that year, as a result of a growing consciousness of the need for a merchant fleet controlled by Nigerians, the Nigerian National Line was established. Today this line owns a number of vessels, charters others, and is a member of the West African Lines Conference which consists of over ten shipping companies.

Phenomenal growth in maritime activities has taken place in Eastern Nigeria over the last half-century. From small ports of entry such as Brass, Buguma, Degema, Bonny and Calabar, with total tonnage movements measured in thousands each year, a modern seaport fully equipped for mechanised handling of cargoes has emerged – Port Harcourt – while Calabar and Bonny are also reviving after years of stagnation and decline. Port Harcourt, forty-one miles from the ocean up the Bonny River, was originally planned as a port terminal for the eastern branch of the railway. The rapid expansion of Nigeria's overseas trade following the Second World War placed a severe strain on the port's facilities, which consisted of four deep-water berths each served by a transit shed and ancillary dumping area, a coal jetty and a lighter berth. The wharf extension scheme completed in 1960 provided three new berths, also transit sheds, offices, road and rail approaches and a modern marshalling yard. Further extensions to the quay and modernisation of facilities are under way.

Port Harcourt handles traffic for the Eastern Provinces, the north-eastern area of Northern Nigeria, and transit traffic to and from the Chad Republic. The new wharf extensions are timed to coincide with the growth of traffic on the 400-mile railway extension from Kuru to Maiduguri since the greater part of the new traffic generated from that area is destined to move to Port Harcourt. The mounting importance of Port Harcourt as an

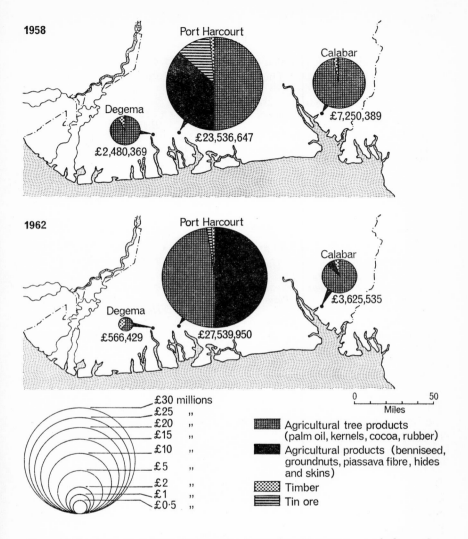

1958

Port Harcourt

Calabar

£7,250,389

Degema

£2,480,369

£23,536,647

1962

Port Harcourt

Calabar

£3,625,535

Degema

£566,429

£27,539,950

£30 millions
£25 "
£20 "
£15 "
£10 "
£5 "
£2 "
£1 "
£0·5 "

0   50
  Miles

▓ Agricultural tree products
   (palm oil, kernels, cocoa, rubber)

■ Agricultural products (benniseed,
   groundnuts, piassava fibre, hides
   and skins)

▒ Timber

≡ Tin ore

FIG. 17.3 *Principal domestic exports through main ports – excluding oil*

all-Nigerian port and its depressant effect upon other Eastern Nigerian harbours is revealed in Tables 17.4 and 17.5. More specific information on the nature and value of exports from Eastern Nigerian ports is indicated in Fig. 17.3.

TABLE 17.4

Foreign trade handled at Nigerian ports: Cargo loaded[1]
Thousands of tons

Year	Total	Lagos	Mid-Western ports			Eastern Nigerian ports		
			Burutu	Warri	Sapele	Port Harcourt	Calabar	Degema
1953	1,620	716	66	29	209	402	117	81
1960	2,779	955	81	51	346	1,190	95	61
1963	6,095	1,217	88	98	386	4,185	83	38

[1] Nigeria, Federal Office of Statistics, *Abstract of Statistics 1964*, Table 6.9, p. 55.

TABLE 17.5

Foreign trade handled at Nigerian ports: Cargo unloaded[1]
Thousands of tons

Year	Total	Lagos	Mid-Western ports			Eastern Nigerian ports		
			Burutu	Warri	Sapele	Port Harcourt	Calabar	Degema
1953	1,506	1,047	44	37	40	282	54	2
1960	3,094	2,172	59	66	104	652	36	5
1963	2,960	2,090	30	39	36	751	14	—

[1] Nigeria, Federal Office of Statistics, *Abstract of Statistics 1964*, Table 6.10, p. 55.

The decline of Calabar, as indicated earlier, may be expected to cease in the immediate future with the opening-up of the trans-Cross River area by road.

The resuscitation of Bonny as an 'oil' port was also well under way prior to the outbreak of the Civil War. This town at the mouth of the Bonny River was once the headquarters of the old Oil Rivers Protectorate and a centre for the export of palm oil. It declined in importance with the completion of the railway, which could approach no nearer to the sea than Port Harcourt. With the establishment of the Shell–BP bulk oil storage plant and shipping terminal at Bonny, and dredging of the bar, Bonny has recaptured much of its former importance as a seaport.

AIRWAYS

The aeronautical facilities necessary for supporting the international and internal air services of Nigeria today are of a relatively complex nature. Formerly, no special facilities were provided, but the growth of air transport has required the establishment of a network of airfields and expensive ancillary services.

The first commercial service between Nigeria and the U.K. was started in 1935 by Imperial Airways, with aircraft travelling via Cairo and Khartoum to Kano and Lagos.[1] With the introduction of this external passenger and airmail service the value of an internal system was realised. In the late 1930s the Public Works Department started construction of over twenty airfields, only a few of which had been completed on the outbreak of hostilities in Europe in 1939. Nevertheless their construction was actively pursued and completed during the war because of their military value.

After the war, international services were reintroduced by British Overseas Airways Corporation (B.O.A.C.), the successor to Imperial Airways, with aircraft now reaching Nigeria via Tripoli and across the Sahara instead of via the lengthy Nile Valley route. In 1946 the West African Airways Corporation was formed for developing both air connections between the British colonies and protectorates of West Africa and internal services. The airfields of Enugu, Port Harcourt and Calabar came into commercial use from this period onwards. At first WAAC used for its services small, twin-engined, passenger-carrying aircraft (De Havilland Doves), but these were soon found to be very costly to operate, with a crew of two and a maximum of six passengers. From the mid-1950s, the DC 3 aircraft became the principal 'work-horse' on internal Nigerian flights.

Since independence, many foreign airlines have joined in the operation of air services to and from Nigeria, among them Pan-American, K.L.M., Air France, Swissair, UAT, Lufthansa, Sabena, Ethiopian Airlines, Ghana Airways. Modern four-engined jet aircraft such as the super VC-10, Boeing 707 and DC 8 make regular scheduled stops at Lagos. The national airline is now officially WAAC (Nigeria) Ltd., but 'Nigerian Airways' is used in all publicity materials, aircraft markings, etc. In 1963 Nigerian Airways put into operation a fleet of five F.27 Fokker Friendship aircraft purchased from Holland.[2] This is one of the most advanced types of aircraft for its size in the world, and offers a superior standard of comfort and efficiency.

---

[1] Anon, 'Civil Aviation', *Nigerian Trade Journal*, Special Independence Issue, 1960 (Lagos, 1960), p. 28.

[2] Anon, 'Progress of Commercial Aviation in Nigeria', *Nigerian Trade Journal*, vol. xii, No. 4 (Oct./Dec. 1964), p. 152.

The introduction of the F.27, which has more passenger and freight space, as well as speed, dependability and comfort than the veteran DC 3, has meant that more Nigerians are travelling by air or shipping goods than ever before. The statistics in Table 17.6 reflect the impact of the new aircraft on the expansion of Nigerian Airways' operations.

TABLE 17.6

Nigerian Airways. Passengers and freight carried, 1962–3[1]

	1962	1963	
Passengers carried	66,448	94,754	
Cargo and excess baggage	685,836	939,030	kilograms
Mail	362,866	397,404	,,
Overall loads per mile	3,874,349	6,307,296	,,

[1] Ibid., p. 153.

Also operating in Eastern Nigeria are a number of charter service companies such as Aero-Contractors of Nigeria Ltd. and Pan African Airlines (Nigeria) Ltd. A chartered helicopter service – Bristow Helicopters Ltd. – for reconnaissance work and VIP travel is utilised by Shell–BP. There is a distinct possibility for the use of hovercraft to serve the creeks, the Delta area and some of the towns on the Niger and the Benue.

18 Development Planning prior to the Civil War

FREQUENT references have been made in previous chapters to the first Eastern Nigeria Development Plan (1962–8) which in turn was a part of the National Development Plan for all Nigeria.[1] The broad objectives of the plan and the progress achieved by 1966 are reviewed in this final chapter of our study. Before the outbreak of hostilities in Nigeria a second 'medium-term' National Development Plan was being prepared, to come into effect at the termination of the present plan and to last for five years (1968–73). An overall 'perspective development' plan covering the period 1968–88 was also being formulated.[2]

The principal feature of the 1962–8 Development Plan was its 'gearing for growth': growth not in the sense of achieving spectacular, immediate or even short-term results, but rather in the sense of laying a solid and enduring foundation for future expansion – an essential prerequisite in Eastern Nigeria's evolution towards a self-sustaining economy.

The prime objective of the plan was to achieve and maintain the highest possible rate of increase in the standard of living of the people. By initiating a series of development projects both in the public and private sectors of the economy, it was intended to maintain and if possible surpass an annual growth rate of four per cent of the Gross Domestic Product (G.D.P.). For this purpose an aggregate capital expenditure of £75,192,000 was envisaged by the Eastern Nigerian Government when the plan was launched in 1962 (Table 18.1).

With over seventy-five per cent of the population of Eastern Nigeria dependent upon agriculture and over fifty per cent of the region's output in the agricultural sector alone, it follows that no serious improvement in the pace of overall progress can be made unless strenuous efforts are made

[1] Nigeria, Federal Ministry of Economic Development, *National Development Plan 1962–68* (Lagos, 1962), chapter 8, Eastern Nigeria Development Programme, pp. 195–266.
[2] *West Africa*, no. 2556 (28 May 1966), p. 606.

to raise productivity in agriculture. Such increases are necessary not only in exports but also in food crops. The aim is to make the Eastern Provinces self-sufficient in food and also to raise the protein content of the diet. One of the main objectives of the plan was therefore the modernisation of agricultural methods through the adoption of improved techniques, intensified agricultural education and changes in land tenure.

TABLE 18.1

Capital and recurrent outlays in the development plan for Eastern Nigeria, 1962–8 (£000)[1]

Sector	Capital	% of total	Recurrent	Total	% of total
1. Primary production (agriculture, forestry, fishing)	30,361	40	6,460	36,821	34
2. Trade and industry	12,930	17	588	13,518	12
3. Transport	8,850	12	1,350	10,200	9
4. Education	8,805	12	21,091	29,896	28
5. Water (urban and rural)	5,100	7	1,100	6,200	6
6. Town and country planning	3,306	4	275	3,581	3
7. General government	2,067	3	382	2,449	2
8. Health	1,819	2	1,381	3,200	3
9. Electricity	600	1	—	600	1
10. Social welfare	534	1	724	1,258	1
11. Information	450	1	190	640	1
12. Justice	250	*	190	440	*
13. Financial obligations	120	*	—	120	*
TOTAL	75,192	100	33,731	108,923	100

* Less than 0·5 per cent.

[1] Eastern Nigeria, *Eastern Nigeria Development Plan 1962–68* (Enugu, 1962), Table ii, p. 13.

It is appreciated that even a steep increase in agricultural productivity cannot absorb the entire annual increase of the working population. Hence the second objective was to provide wider employment opportunities by diversification of the economy. An increased rate of industrialisation was to be stimulated – industries based on labour-intensive techniques would be especially welcome – and there would be substantial investment in building and other construction work.[1] The stage of development reached

[1] Ibid., p. 8.

in agriculture and industry by 1966 was due largely to private enterprise: export crops had been developed on a peasant basis and manufacturing had grown largely through private investment. The third objective was therefore 'to strengthen the private sector as much as possible, to continue to provide assistance – administrative, financial, and technical – for the promotion of rationalised agriculture and for the development of manufacturing industries.'[1]

It was recognised that shortages of skilled manpower might prove a limiting factor in achieving the first three objectives. A fourth aim was therefore to raise the level of technology in Eastern Nigeria by the encouragement of technical training and science education. The strong Science, Agricultural and Engineering Faculties of the University of Nigeria, Nsukka, symbolised this significant trend in higher education.

The Government was also aware of its responsibility for aiding the advancement of socially- and economically-retarded areas within the Eastern Provinces. 'It would fail in its obligations if it were to neglect the needs of special areas or to concentrate development in a few areas simply because they already have the basic facilities installed.'[2] An important objective of the 1962–8 plan was therefore to reduce the disparities in levels of development in different parts of Eastern Nigeria. Areas such as the Cross River Basin and the Niger Delta (via the Niger Delta Development Board) were to receive special consideration.

Finally, it was the aim of the Six-Year Development Plan to evoke as wide a general participation by the people as possible in development growth. The spirit of self-help is particularly strong in Eastern Nigeria; one of the objectives of the Plan was to harness this spirit through enlarged Community Development programmes.[3]

A key feature of the plan was its flexibility. It was realistically appreciated that not all sectors for which provisions had been made would reach the same degree of elaboration or come to full fruition within the plan period. In addition, experience gained in the first years of the plan enabled its architects to review the various projects and to make adjustments as necessary. As sizeable external assistance was needed to implement the plan in full, there was a continuing obligation to review the allocations in light of the assistance received for specific projects.

The first progress report on the National and Regional Plans was published in 1965.[4] The report indicated that, by and large, a good start

[1] Ibid., p. 8.   [2] Ibid., p. 9.   [3] Ibid.
[4] Nigeria, Federal Ministry of Economic Development, *National Development Plan. Progress Report 1964* (Lagos, 1965), chap. 4, app. B, Eastern Nigeria Government Programme, pp. 136–67.

had been made in the new approach towards resolving the development problems of the nation. In Eastern Nigeria, expenditure in all sectors over the first two years amounted to £20·2 million. Slightly more than half of this went into the development of agriculture and industry, the two priority sectors of the plan. The development of tree crops through rehabilitation of existing groves and establishment of modern plantations led to the planting of over 29,000 acres of oil palm, nearly 23,000 acres of rubber and some 13,000 acres of cocoa. Much attention was given to the development of the cattle and pig industries, poultry farming and the fishing industry, including fish ponds, freshwater and deep-sea fishing. There was considerable capital expenditure on expanding facilities at the Umudike Agricultural Research Centre.

From 1962 to 1964, over £4,900,000 was invested in various industrial projects such as textiles, flour mill, plastics, shoes, oil refining, ceramics, brewing, glass manufacture, cement works, asbestos roofing sheets and water pipes, vehicle assembly, light engineering works and tyre manufacture. Most of these investments were in partnership with private foreign concerns. Substantial progress was made on all major projects in the road development programme, with a total expenditure of over £3 million on road construction and maintenance. Agricultural roads, including farm-settlement and plantation-access roads, received particular attention.

Several water-supply schemes were completed and commissioned during the first two years of the plan, raising the total capacity of rural water supply from 1·3 million gallons daily before the plan to 6·7 million gallons a day. The proposed capacity at the end of the plan period is 16·5 million gallons daily.[1] Although the progress report predicted no overall increase in the cost of Eastern Nigeria's programme, a slight shift of planned investment from one sector to another has occurred (Table 18.2).

By the end of 1965 it was apparent that the prime objective of the plan was being met. An annual growth rate of over four per cent had been achieved and, for the first time in ten years, there was every indication that the country as a whole had a balance of payments surplus. The prospects for a continuation of this impressive growth rate and the economic expansion of Eastern Nigeria were on the whole good. The petroleum industry was providing handsome returns, with crude oil the largest single item on Nigeria's export list. In 1965 oil production increased 126 per cent over 1964, and exports amounted to £68 million in value. Oil exports to the value of over £100 million were reached in 1966. However, the benefit to the economy of this vital industry should not be exaggerated. Remittances

[1] Ibid., p. 165.

TABLE 18.2

Revised costs of the Eastern Nigeria development programme (1964) (£000)[1]

Sector	Original cost of projects	% of Total	Revised cost of projects	% of Total	Increase (+) or decrease (−) in cost of projects
I. *Economic*					
Primary production	30,361	40·4	27,896	37·1	− 2,645
Trade and industry	12,930	17·2	11,436	15·2	− 1,494
Electricity	600	0·8	758	1·0	+ 158
Transport	8,850	11·8	10,350	13·8	+1,500
TOTAL	52,741	70·2	50,440	67·1	− 2,301
II. *Social Overheads*					
Water supplies	5,100	6·8	5,100	6·8	—
Education	8,805	11·7	8,085	10·8	− 720
Health	1,819	2·3	1,856	2·4	+ 37
Town and country planning	3,306	4·4	5,163	6·9	+1,857
Social welfare	534	0·7	534	0·7	—
Information	450	0·6	1,187	1·6	+ 737
TOTAL	20,014	26·5	21,925	29·2	+1,911
III. *General Administration*					
Judicial	250	0·3	250	0·3	—
General	2,067	2·8	2,407	3·2	+ 340
TOTAL	2,317	3·1	2,657	3·5	+ 340
IV. *Financial Obligations*					
Financial obligations	120	0·2	170	0·2	+ 50
GRAND TOTAL	75,192	100·0	75,192	100·0	0

[1] Ibid., Table 4.3, p. 63.

of profits and outflow of capital funds meant that the balance of payments benefited less from the oil industry than the gross value of oil exports would suggest.[1] Although oil discoveries and exports were encouraging, the very size of the territory and population of Nigeria means that the benefits accruing to Kuwait or Libya, for example, cannot possibly be looked for in Nigeria. Nevertheless, the advantages stemming from the oilfields of the Delta are considerable, and may be counted on to contribute substantially to the future development of the Eastern Provinces.

[1] Anon, 'Nigeria on the Move', *West Africa*, no. 2544, 5 Mar. 1966, p. 265.

The size of the population, while in many ways a handicap to development planning, does mean a large potential domestic market for Eastern Nigerian products; with rising *per caput* incomes, industrialisation was being greatly aided prior to the civil war. It makes sense with a large internal market to go in for 'import substitute' industries in a big way as was happening in the Eastern Provinces. Such an emphasis had a beneficial effect on the trade balance both in the East and in the nation at large.

Despite the political disturbances which led to the establishment of military rule in January 1966, Eastern Nigeria was still an attractive field for capital investment before the outbreak of hostilities in 1967. The encouragement given to foreign investment by the former civil Government was continued under the National Military Government. An indication of confidence in the future of Nigeria and the Eastern Provinces was provided by substantial increases in financial backing awarded by the Consultative Group (a consortium of the major aid-giving countries under the auspices of the World Bank) in February 1966.[1] These investments provided sufficient external finance to allow Nigeria to carry out the remainder of the first Development Plan. Equally important, the countries involved – the U.S.A., U.K., Germany, Netherlands, Italy, Belgium, Canada and Japan – made it clear that they would give full support to the next Development Plan. Nigeria's acceptance as an associated member of the European Economic Community (the Common Market) in July 1966 was also evidence of international confidence in the nation's future.

However it would be wrong to gloss over or to ignore some of the problems which faced Eastern Nigeria in her struggle for economic advancement; some fundamental rethinking will be necessary if these problems are to be successfully overcome, during the reconstruction period following the end of the war.

In industry, certain projects had been ill-conceived and the whole approach to industrialisation required revision. For example, it is evident that 'substitution of imports doesn't in itself create wealth or give jobs: truly viable industries have to be broadly based – the bulk of their raw materials must be available on the spot, and they must be linked to each other'.[2] In 1966, few of the existing industries in the East could pass these tests.

The experience of several industries had been disappointing, notably the Umuahia Ceramics factory, the Golden Guinea brewery, the Pepsi-

---

[1] Anon, 'Backing Nigeria's Future', *West Africa*, no. 2542, 19 Feb. 1966, p. 197.

[2] Anon, 'New Thinking in Eastern Nigeria: 1', *West Africa*, no. 2654, 25 Dec. 1965, p. 1457.

Cola Bottling Plant and the Port Harcourt Glass Factory. These ENDC-sponsored and managed plants tended to be over-staffed with non-productive and unqualified personnel and their efficient operation had suffered accordingly. There were also disappointments with their technical operations which, in the case of the ceramics factory and the brewery, had been in the hands of expatriate technicians.

Another basic problem was 'expatriate machine selling'. Under this arrangement, foreign firms offered to provide machinery for a new industry lock, stock and barrel, without assuming any responsibility for the ultimate success of the factory. Indeed, the machinery may have been obsolete and valueless. The foreign companies may have had a small shareholding but if the industry failed to live up to expectations the losses were borne by Nigerians; and those experienced by the expatriate suppliers of the plant were more than offset by the profits made by selling the machinery.

A further problem arose from high overheads due to small plants, heavy costs of establishment and pre-production expenses. These included housing for expatriate staff and other benefits, apprenticeship training programmes, importation of machinery and replacement of spares. Essential parts invariably had to be sent abroad for servicing, with subsequent delays in transit and through customs.

Another universal problem stemmed from labour and management relationships. There were innumerable cases of irresponsible leadership on the part of the unions as well as unnecessary intransigence on the part of management. Constant upheavals and threats of strikes are not conducive to increased productivity and efficiency in industrial establishments. Improving the level of management, steering clear of 'machine-selling projects' in favour of more broadly-based industries such as pulp and paper, kernel-cracking and soapmaking, and fostering the further development of small, labour-intensive industries: these are among the measures which need promoting in order to advance the industrial 'revolution' in Eastern Nigeria.

In the rural sector, the imposition of land reform schemes and agricultural betterment projects 'from the top', i.e. the Ministries, clearly has its limitations. Such planning 'takes a long time to percolate down to the villages, and between the planners and the villages lie a host of obstacles, including inertia, corruption and ignorance'.[1] The large-scale farm settlements and plantation projects can only make a limited impact on the rural populace during their formative years. More 'grass-roots' community projects at the village level, taking stock of the total resources and poten-

[1] Anon., 'New Thinking in Eastern Nigeria: 2', *West Africa*, no. 2535, 1 Jan. 1966, p. 7.

tialities of each village, and leading to 'development from the bottom rather than from the top' may be the answer to the problems of sounder land-use management in the rural areas.

In sum, the sustained economic and social advancement of Eastern Nigeria will require a continued, systematic exploration of the region's natural and human resources for rational development. Transformation of a traditional and partially Westernised society into a modern, diversified and virtually self-sustaining socio-economic system is the ultimate goal. From our geographer's vantage-point we have surveyed objectively the essential ingredients of earth and man from which this new society must be forged. We may take comfort from the lorry slogan: 'No Condition is Permanent'. With wisdom in the framing of Eastern Nigeria's future policies, and discipline and dedication on the part of all Easterners in their execution, there is no reason why the endowments of nature cannot be developed to support a lasting improvement in living standards. It is on the basis of such a faith that this geographical study of Eastern Nigeria has been written.

APPENDIX

The Succession of Geologic
Strata in Eastern Nigeria

With an Outline of their Lithology and
Related Topography

Period	Age	Formation[1]	Max. thickness (ft)
Quaternary	Recent (Holocene)	Alluvium	—
	Recent (Pleistocene)	Delta Formation	10,000 +
Tertiary	Pliocene	Coastal Plains Sands (Benin Formation)	5,000 +
	—Unconformity— Miocene	Lignite Formation (Ogwashi-Asaba Formation)	1,700 +
	Eocene	Bende-Ameki Formation (Ameki Formation)	4,800
	Palaeocene-Eocene	Imo-Clay Shales Formation (Imo Shale Formation)	4,000 +
	Palaeocene	Upper Coal Measures Formation (Nsukka Formation)	1,100 +
Cretaceous	Maastrichtian		
	Maastrichtian	Falsebedded Sandstone Formation (Ajali Formation)	1,100 +
	Maastrichtian	Lower Coal Measures Formation (Mamu Formation)	1,300
	Maastrichtian-Campian (?)	Enugu Shales	2,500 +
	Maastrichtian-Campian (?)	Awgu Sandstone	2,000
		Nkporo Shales	1,500
	—Unconformity— Coniacian-Turonian (?)	Awgu-Ndeaboh Shale Formation (Awgu Shale Formation)	3,000
	Turonian	Eze-Aku Shale Formation	2,000
	Albian-Cenomanian	Asu River Group	6,000 +
Palaeozoic (?) Pre-Cambrian Archaean (?)	—Unconformity—	Basement Complex	?

[1] Names in brackets are those given in Reyment (1965) and are not in general current use.

Lithology	Origin of materials[1]	Topography
...consolidated sands with subordinate gravels, clays, ...s.		Flat valleys of Niger and Cross Rivers Low-lying land, tidal creeks
...consolidated dominantly yellow and white sands, ...casionally pebbly with clay beds of various colours.	C	Gently sloping and undulating plain, more hilly in north-west.
...oss-bedded sandstones, often coarse; carbonaceous ...dstones and lignite; variegated clays, often sandy ...d usually plastic.	C	Rolling, undulating lowland.
...ne to coarse sandstone, generally red, white, yellow; ...ley sandstones, shales, and sandy shales, mudstones ...d rare thin sandy limestones.	M	Dissected plateau, ridges and valleys
...ue, dark grey shales with occasional bands of clay ...nstone and thin argillaceous sandstones. Intra-...mational sandstones at several horizons.	M	Lowlands and ridges
...hite to grey, coarse to medium sandstones, carbona-...ous shales, sandy shales, thin coals, occasional lime-...ne.	C	Minor cuesta, outlier hills (buttes) on Nsukka-Udi Plateau
...ne to coarse-grained, cross-bedded friable sand-...nes, sometimes poorly consolidated, with occasion-...bands of white or pale grey shade.	C	Plateau in north, cuesta in south. Escarpment.
...ne to medium-grained, white to grey sandstones, ...ley sandstones and sandy shales; grey mudstones ...d shales. Subordinate calcareous shales and coals.	C	Escarpment (lower slopes)
...uish-grey to dark grey mudstones and shales with ...ds of white sandstone and striped sandy shale. ...casional limestone beds in the lower part.	M	
...ne to coarse-grained massive sandstones, locally ...oss-bedded, with some pebbly beds and subordinate ...nds of siltstone and carbonaceous shale. Shales and ...nestone at base.	M	Escarpment (lower slopes) and immediate lowlands at foot of scarp.
...rk shales and mudstones with occasional thin beds ...sandy shale and sandstone; very thick lenses of ...ndstone present in Afikpo area with associated ...rbonaceous shales and thin coals.	M	
...uish-grey, well-bedded shales with subordinate ...lly limestones and sandstones. Strong development ...sandstone in Afikpo area.	M	
...uggy, calcareous shales and mudstones and sub-...dinate shelly limestones and sandstones.	M	Cross River Plain, an extensive lowland with minor undulations formed by more resistant sand-stones and limestones. Occasional prominences from volcanic materials.
...uish-grey to olive-brown shales and sandy shales, ...e-grained micaceous sandstones and dense blue ...d grey limestones.	M	
...neisses, schists, granites and associated crystalline ...cks.		Mountainous terrain of the Eastern Highlands, with planated margins.

[1] C Essentially Continental; M Essentially Marine.

Bibliography

I. PERIODICALS (CONTINUING SOURCES)

The following English-language periodic publications feature up-to-date contributions on African, West African, Nigerian and Eastern Nigerian geography. Reference should be made to these sources for the latest data on developments in the continent and the region of study featured in this book.

A. AFRICA, WEST AFRICA

Africa (quarterly)
Africa Report (monthly)
African Soils (quarterly)
Annals of the Association of American Geographers (quarterly)
Economic Geography (quarterly)
Focus (monthly)
Geographical Journal (quarterly)
Geographical Magazine (monthly)
Geographical Review (quarterly)
Geography (quarterly)
Journal of Tropical Geography (quarterly)
National Geographic Magazine (monthly)
Outlook on Agriculture (quarterly)
Scottish Geographical Magazine (quarterly)
Tijdschrift voor Economische en Sociale Geografie (quarterly, contributions in English)
Tropical Agriculture (quarterly)
West Africa (weekly)
West African Directory (annual)

B. NIGERIA, EASTERN NIGERIA

Agro-Meteorological Bulletin (monthly): Nigerian Met. Service

Annual Abstract of Statistics: Nigeria (Fed.), Office of Statistics

Annual Report: Eastern Nigeria Development Corporation

Annual Report: Eastern Nigeria Marketing Board

Annual Report: Eastern Nigeria, Ministry of Agriculture (Agricultural Division)

Annual Report: Eastern Nigeria, Ministry of Agriculture (Forestry Division)

Annual Report: Eastern Nigeria, Ministry of Commerce

Annual Report: Nigerian Coal Corporation

Annual Report: Nigerian Geological Survey

Annual Report: Nigerian Railway Corporation

Crop and Weather Report (monthly): Eastern Nigeria, Ministry of Agriculture

Development (quarterly): Eastern Nigeria Development Corporation

Eastern Nigeria (monthly): Eastern Nigeria, Ministry of Information

Extension Newsletter (monthly): Eastern Nigeria, Ministry of Agriculture

Farm and Forest (quarterly: ceased publication)

Handbook of Commerce and Industry (occasional): Nigeria (Fed.), Ministry of Information

Journal of the West African Science Association (quarterly)

Journal of the Historical Society of Nigeria (quarterly)

Nigeria Magazine (quarterly): Nigeria (Fed.), Ministry of Information

Nigerian Field (quarterly): Nigerian Field Society

Nigerian Geographical Journal (half-yearly): Nigerian Geographical Society .

Nigerian Grower and Producer (and West African Farmer) (quarterly): University of Ibadan

Nigerian Journal of Economics and Social Studies (quarterly): Nigerian Institute of Social and Economic Research, University of Ibadan

Nigerian Opinion (monthly)

Nigerian Scientist (occasional)

Nigerian Trade Journal (quarterly): Nigeria (Fed.), Ministry of Commerce and Industries

Overseas Review (monthly): Barclays D.C.O.

Overseas Survey (annual): Barclays D.C.O.

Shell–BP Bulletin (formerly *Oil Search Bulletin*): (bi-monthly, occasionally monthly): Shell–BP Petroleum Development Company of Nigeria Ltd.

Statistical Digest of Eastern Nigeria (annual): Eastern Nigeria, Ministry of Economic Planning

Trade and Industrial Bulletin (monthly): Eastern Nigeria, Ministry of Commerce

University Geographer (annual): Geography Society, University of Ibadan

II. GENERAL REFERENCES

A. AFRICA, WEST AFRICA

Books, Monographs, Pamphlets, etc.

Among recent publications dealing with the broad physical and human geographical aspects of the African continent are:

Barbour, K. M., *Population in Africa* (Ibadan, 1963).
— and Mansell Prothero, R. (eds.), *Essays on African Population* (London, 1961).
Bascom, W. R., and Herskovits, M. J., *Continuity and Change in African Cultures* (Chicago, 1959).
Caldwell, J. C., and Okonjo, C. (eds.), *The Population of Tropical Africa* (London, 1968).
Carlson, L., *Africa's Lands and Nations* (N.Y., 1967).
DeBlij, H., *A Geography of Subsaharan Africa* (Chicago, 1964).
Fitzgerald, W., *Africa, a Social, Economic and Political Geography* (London, 1967).
Forde, D. (ed.), *African Worlds* (London, 1954).
Fordham, P., *The Geography of African Affairs* (Pelican, London, 1965).
Gibbs, J. L. (ed.), *Peoples of Africa* (N.Y., 1965).
Grove, A. T., *Africa South of the Sahara* (London, 1967).
Hailey, Lord, *An African Survey* (London, 1956).
Hance, W. A., *The Geography of Modern Africa* (New York, 1964).
Hargreaves, J. F., *Prelude to the Partition of West Africa* (N.Y., 1963).
Harrison Church, R. J., *West Africa. A Study of the Environment and of Man's Use of It* (London, 1963).
— *Environment and Politics in West Africa* (Princeton, N.J., 1963).
— *et al.*, *Africa and the Islands* (London, 1964).
Hatch, J., *A History of Postwar Africa* (N.Y., 1965).
Hazelwood, A., *The Economy of Africa* (London, 1961).
Herskovits, M. J., *The Human Factor in Changing Africa* (New York, 1963).
Hodder, B., and Harris, D., *Africa in Transition* (London, 1967).

Hodgson, R. D., and Stoneman, E. A., *The Changing Map of Africa* (N.Y., 1963).

Hughes, J., *The New Face of Africa* (London, 1962).

Hunter, G., *The New Societies of Tropical Africa* (London, 1962).

Irvine, F. R., *A Text Book of West African Agriculture* (London, 1953).

Jarrett, H. R., *A Geography of West Africa* (London, 1960).

Johnson, B. F., *The Staple Food Economies of Western Tropical Africa* (Stanford, 1958).

Kimble, G. H. T., *Tropical Africa*. 2 vols. (New York, 1961).

Legum, C. (ed.), *Africa: A Handbook to the Continent* (N.Y., 1966).

Mair, L. P., *Studies in Applied Anthropology* (London, 1957).

Mountjoy, A. B., and Embleton, C., *Africa* (London, 1967).

Murdoch, G. P., *Africa. Its Peoples and their Cultural History* (New York, 1959).

Oxford Economic Atlas: Africa (Oxford, 1965).

Ottenberg, S. and P., *Cultures and Societies of Africa* (New York, 1960).

Perham, M., *African Outline* (London, 1965).

Southall, A. (ed.), *Social Change in Modern Africa* (London, 1961).

Stamp, L. D., *Africa, a Study in Tropical Development* (London, 1966).

The Times Atlas of the World, vol. iv (London, 1956).

UNESCO, *A Review of the Natural Resources of the African Continent* (Paris, 1963).

Worthington, E. B., *Science in the Development of Africa* (London, 1958).

Wraith, R., *Local Government in West Africa* (London, 1964).

B. NIGERIA, EASTERN NIGERIA

Books, Monographs, Pamphlets, etc.

Ademoyega, W., *The Federation of Nigeria* (London, 1962).

Aluko, S., *The Problems of Self-Government for Nigeria: a Critical Analysis* (Ilfracombe, England, 1955).

Anon, *Directory of the Federation of Nigeria* (The Diplomatic Press and Publishing Co., London, 1962).

Awa, E. O., *Voting Behaviour and Attitudes of Eastern Nigerians* (Ibadan, 1961).

Awolowo, Chief O., *Thoughts on Nigerian Constitution* (Ibadan, 1966).

Azikiwe, N., *Nigeria in World Politics* (London, 1959).

Bretton, H. L., *Power and Stability in Nigeria*. The Politics of Decolonization (New York, 1962).

Buchanan, K. M. and Pugh, J. C., *Land and People in Nigeria* (London, 1958).

Coleman, J. S., *Nigeria: Background to Nationalism* (Berkeley, 1958).

Collis, R., *A Doctor's Nigeria* (London, 1961).

Dickson, M., *New Nigerians* (Chicago, 1963).

Dudley, B. J. (ed.), *Nigeria: Crisis and Criticism* (Ibadan, 1966).

Eastern Nigeria, Ministry of Information, *Eastern Nigeria*. Independence Edition (Enugu, 1960).

— Ministry of Economic Planning, *Statistical Digest of Eastern Nigeria*. 1st ed. Official Document No. 22 of 1963 (Enugu, 1963).

— — *Statistical Digest of Eastern Nigeria*. 2nd ed. Official Document No. 24 of 1965 (Enugu, 1965).

— Ministry of Works, *Master Plan for Urban and Rural Water Supply* (Enugu, 1962). Submitted by Eastern Nigeria Water Planning and Construction Ltd. Prepared by TAHAL (Water Planning) Ltd, Tel-Aviv, Israel.

Ezera, K., *Constitutional Developments in Nigeria* (Cambridge, 1960).

Forde, D. and Scott, R., *The Native Economies of Nigeria* (London, 1946).

Gwam, L. C., *A Handlist of Nigerian Official Publications* (provisional) 2 vols. (Ibadan, 1961).

— *A Preliminary Index to the Intelligence Reports in the Nigerian Secretariat Record Group*. National Archives (Ibadan, 1961).

Harris, P. J., *Local Government in Southern Nigeria*. Manual of Procedure and Text of the Laws (Cambridge, 1957).

Iloeje, N. P., *A New Geography of Nigeria* (Lagos, 1965).

Iyanam, U. I., 'The Human Geography of Oron County, Eket Division'. University of Ibadan, Original Essays in Geography (Ibadan, 1962). Manuscript.

Jennings, J. and Oduah, S., *A Geography of Eastern Nigeria* (London, 1966).

Mbah, V. C. J., *A Preliminary Index to the Intelligence Reports in the Enugu Secretariat Group*. National Archives (Enugu, 1962).

Menakaya, J. C. and Floyd, B. N. (eds.), *Junior Atlas for Eastern Nigeria* (London, 1965).

Ndem, E. B. E., *Ibos in Contemporary Nigerian Politics*. A Study in Group Conflict (Onitsha, 1961).

Nigeria, Director of Surveys, *Gazetteer of Place Names on 1 : 500,000 Map of Nigeria* (Lagos, 1949).

— Ministry of Commerce and Industry. *Nigerian Trade Journal*. Special Independence Issue (Sept. 1960).

— Ministry of Information, *Nigeria 1960*. Special Independence Issue of *Nigeria Magazine*; M. Crowder (ed.) (Oct. 1960).

Oboli, H. O. N., *A Sketch-Map Atlas of Nigeria* (London, 1960).

Omali, D. O., *A Nigerian Villager in Two Worlds* (London, 1965).

Pedler, F. J., *Economic Geography of West Africa* (London, 1955).

Perkins, W. A., and Stembridge, J. H., *Nigeria: A Descriptive Geography* (London, 1966).

Pugh, J. C., and Perry, A. C., *Short Geography of West Africa* (London, 1960).

Quinn-Young, C. T., and Herdman, T., *Geography of Nigeria* (London, 1964).

Royal Institute of International Affairs, *Nigeria: the Political and Economic Background* (London, 1960).

Sircar, P. K. (ed.), *Nsukka Division. A Geographic Appraisal*, University of Nigeria (Nsukka, 1965). Cyclostyled.

Sklar, R. L., *Nigerian Political Parties: Power in an Emergent African Nation* (Princeton, N.J., 1963).

Stapleton, G. B., *The Wealth of Nigeria* (London, 1958).

Tilman, R. O., and Cole, T. (eds.), *The Nigerian Political Scene* (Durham, N.C., 1962).

Watson, G. D., *A Human Geography of Nigeria* (London, 1960).

U.K., *Report of the Commission of Enquiry into the Disorders in the Eastern Provinces of Nigeria, Nov. 1949* Col. No. 256 (London, 1950).

— *Willink Commission on the Problems of Nigerian Minorities* (Lagos, 1958).

Periodicals

Adebanjo, T., *et al.*, 'Six Views of the Nigerian War', *Africa Report*, vol. 13, no. 2 (Feb. 1968).

Anene, J. C., 'The Nigeria–Southern Cameroons Boundary', *Journal of the Historical Society of Nigeria*, vol. ii, no. 1 (Dec. 1961), 186–95.

Buchanan, K. M., 'Nigeria – Largest Remaining British Colony', *Economic Geography*, vol. xxviii (1952), 302–22.

Floyd, B. N., 'The Federation of Nigeria', *Focus* (Nov. 1964).

Frodin, R., 'A Note on Nigeria', *Reports Service*, American Universities Field Staff, West Africa Series, vol. iv, no. 6 (Nigeria) (Aug. 1961).

Kitson, A. E., 'Southern Nigeria. Some Considerations of its Structure, People and Natural History', *Geographical Journal*, vol. xli, no. 1 (Jan. 1913), 16–38.

Mitchel, N. C., 'Nigeria', *Focus* (Mar. 1954).

Ogunsheye, A., 'Nigeria's Economy', *Nigeria Magazine*, no. 66 (Oct. 1960), 12–19.

Prescott, J. R. V., 'The Geographical Basis of Nigerian Federation', *Nigerian Geographical Journal*, vol. ii, no. 1 (June 1958), 1–13.

— 'The Evolution of Nigeria's Boundaries', *Nigerian Geographical Journal*, vol. ii, no. 2 (Mar. 1959), 80–104.

— 'Nigeria's Regional Boundary Problems', *Geographical Review*, vol. xlix (1959), 485–505.

III. HISTORICAL ASPECTS

Books, Monographs, Pamphlets, etc.

Ademoyega, 'Wale, *The Federation of Nigeria*, from Earliest Times to Independence (London, 1962).

Burns, A., *History of Nigeria* (London, 1948).

Crowder, M., *The Story of Nigeria* (London, 1962).

Dike, K. O., *Trade and Politics in the Niger Delta 1832–1885* (Oxford, 1956).

English, M. C., *An Outline of Nigerian History* (London, 1960).

Fage, W., *An Atlas of African History* (London, 1958).

Flint, J. E., *Sir George Goldie and the Making of Nigeria* (Oxford, 1962).

Forde, D., *Efik Traders of Old Calabar* (London, 1956).

Hodgkin, T., *Nigerian Perspectives*: an Historical Anthology (Oxford, 1960).

Jones, G. I., *The Trading States of the Oil Rivers*: a Study of Political Development in Eastern Nigeria (London, 1963).

Perham, M., *Native Administration in Nigeria* (London, 1937).

Periodicals

Hartle, D. D., 'Archaeology in Eastern Nigeria', *Nigeria Magazine*, no. 93 (June 1967), 134–43.

IV. SOCIOLOGICAL ASPECTS

Books, Monographs, Pamphlets, etc.

Basden, G. T., *Niger Ibos* (London, 1921).

Chubb, L. T., *Ibo Land Tenure* (Ibadan, 1961).

Eastern Nigeria, Ministry of Internal Affairs, *Community Development in Eastern Nigeria*. Official Document No. 20 of 1962 (Enugu, 1962).

Elias, T. O., *Nigerian Land Law and Custom* (London, 1962).

Eme, J. C. U., 'Sociological Problems Connected with Farm Settlement Schemes', Appendix VI in Eastern Nigeria, Ministry of Agriculture, *Farm Settlement Scheme*. Third Annual Programme Planning Conference. Technical Bulletin No. 4 (Enugu, 1963).

Forde, D., and Jones, G. I., *The Ibo and Ibibio speaking Peoples of Southeastern Nigeria*. International African Institute (London, 1950).

Green, M. M., *Land Tenure in an Ibo Village in Southeastern Nigeria*, London School of Economics, Monographs in Social Anthropology No. 6 (London, 1941).

— *Ibo Village Affairs* (London, 1948).

Inyang, P. E, B., *Language Groups of Eastern Nigeria*, Geography Department, University of Nigeria (Nsukka, 1963). Manuscript.

Jackson, I. C., *Advance in Africa*; a Study of Community Development in Eastern Nigeria (London, 1956).

Leonard, A. G., *The Lower Niger and its Tribes* (London, 1906).

Meek, C. K., *Law and Authority in a Nigerian Tribe* (London, 1937). A Study in Indirect Rule.

— *Land Tenure and Land Administration in Nigeria and the Cameroons*, Col. Research Studies No. 22 (London, 1957).

Obi, S. N. C., *Ibo Law of Property* (London, 1963).

Ottenberg, P., 'The Afikpo Ibo of Eastern Nigeria', in J. L. Gibbs (ed.), *Peoples of Africa* (N.Y., 1965).

Ottenberg, S., 'Ibo Receptivity to Change', in W. R. Bascom and M. J. Herskovits, *Continuity and Change in African Cultures* (Chicago, 1959).

Schwartz, F. A., *Nigeria: the Tribes, the Nation or the Race* (Cambridge, Mass., 1965).

Smythe, H. B., and M. M., *The New Nigerian Elite* (Stanford, 1960).

Talbot, P. A., *Peoples of Southern Nigeria*. 4 vols. (London, 1926).

— *Tribes of the Niger Delta* (London, 1932).

— and Mulhall, H., *The Physical Anthropology of Southern Nigeria*, A Biometric Study in Statistical Method. Occasional Publications of the Cambridge University Museum of Archaeology and Ethnology (Cambridge, 1962).

Periodicals

Alagoa, E. J., 'Ijo Origins and Migrations', *Nigeria Magazine*, no. 92 (Mar. 1967), 47–55.

Anumba, E. O., 'Geographical Significance of Social Institutions in Eastern Nigeria', *The University Geographer* (Ibadan), vol. iii, no. 2 (May 1962), 17–20.

Ardener, E. W., 'Some Ibo Attitudes to Skin Pigmentation', *Man*, vol. ci (1954), 1–3.

Brinkworth, I., 'Nigeria's Cultural Heritage', *Geographical Magazine*, vol. xxxi, no. 9 (Jan. 1959), 425–38.

Forde, C. F., 'Land and Labour in a Cross River Village, South Nigeria', *Geographical Journal*, vol. xc (1937), 24–51.

Harris, J. S., 'Papers on the Economic Aspects of Life among the Ozuitem Ibo', *Africa*, vol. xiv (1943), 12–23.

— 'Some Aspects of the Economies of Sixteen Ibo Individuals', *Africa*, vol. xiv (1944), 302–35.

Jones, G. I., 'Ibo Land Tenure', *Africa*, vol. xix (1949), 309–23.

— 'Dual Organization in Ibo Social Structure', *Africa*, vol. xix (1949), 150–6.

Morgan, W. B., 'The Influence of European Contacts on the Landscape of Southern Nigeria', *Geographical Journal*, vol. cxxv, Pt. 1 (Mar. 1959), 48–64.

Ntukidem, A. E., 'Rural Markets in Nigeria', *University Geographer* (Ibadan), vol. iii, no. 2 (May 1962), 27–29.

Nzekwu, O., 'Ibo People's Costumes', *Nigeria Magazine*, vol. lxxviii (Sept. 1963), 164–75.

Ottenberg, S., 'The Present State of Ibo Studies', *Journal of the Historical Society of Nigeria*, vol. ii, no. 2 (Dec. 1961).

Porter, P. W., 'Environmental Potentials and Economic Opportunities – a Background for Cultural Adaptation', in W. Goldschmidt *et al.*, 'Variation and Adaptability of Culture: a Symposium', *American Anthropologist*, vol. lxviii (1965), 409–20.

Ukeje, L. O., 'Weaving in Akwete', *Nigeria Magazine*, vol. lxxiv (Sept. 1962), 32–41.

V. POPULATION AND SETTLEMENT

Books, Monographs, Pamphlets, etc.

Anugwelem, S. C., *The Urban Growth of Aba*, University of Ibadan, Original Essays in Geography (Ibadan, 1962). Manuscripts.

Emenyi, E. A. H., 'Geography of Oron Town, A River Settlement', University of Ibadan, Original Essays in Geography (Ibadan, 1962). Manuscript.

Karmon, Y., *A Geography of Settlement in Eastern Nigeria*, Scripta Hiero-solymitana 15, Pamphlet No. 2 (Jerusalem, The Magness Press for the Hebrew University, 1966), 90 pp.

Madukaife, G. N., 'The Impact of Migration on Agriculture in Ogidi', Original Essays in Geography, University of Nigeria (Nsukka, 1966). Manuscript.

Mitter, O. K., 'On Birth and Death Rates in Nigeria', Mathematics Department, University of Nigeria (Nsukka, 1965). Manuscript.

Nigeria, Department of Statistics, *Population Census of the Eastern Region of Nigeria* (Lagos, 1953).

— Federal Ministry of Commerce and Industry, *This is Nigeria: Onitsha* (Lagos, n.d.).

Port Harcourt Chamber of Commerce, *Port Harcourt* (Port Harcourt, n.d.).

Wigwe, G. A., 'Trade, Commercial and Industrial Development of Aba', University of Ibadan, Original Essays in Geography (1962). Manuscript.

Periodicals

Allpress, P. L., 'Post-war Settlement and Resettlement in the Eastern Provinces of Nigeria', *Farm and Forest*, vol. vii (1946).

Buchanan, K. M., 'Internal Colonization in Nigeria', *Geographical Review*, vol. xliii (1953), 416–18.

Hair, P. E. H., 'Enugu: An Industrial and Urban Community in East Nigeria, 1914–1953', *Proceedings of the Second Annual Conference, West Africa Institute of Social and Economic Research* (Ibadan, 1953), 143–67.

Jennings, J. H., 'A Population Distribution Map of the Eastern Region of Nigeria', *Geographical Journal*, vol. cxxiii (1957), p. 416.

— 'Enugu – a Geographical Outline', *Nigerian Geographical Journal*, vol. iii, no. 1 (Dec. 1959), 28–38.

Jones, G. I., 'The Method of Obtaining Slaves in the Ibo Country and hence the Effect on Population', *Journal of the Royal Anthropological Institute*, vol. lxxix (1949).

Mabogunje, A., 'The Economic Implication of the Pattern of Urbanisa-tion in Nigeria', *Nigerian Journal of Economic and Social Studies*, vol. vii, no. 1 (Mar. 1965), 9–30.

Morgan, W. B., 'Farming Practice, Settlement Pattern, and Population Density in South-Eastern Nigeria', *Geographical Journal*, vol. cxxi (Sept. 1955), 320–33.

Morgan, W. B., ' The "grassland towns" of the Eastern Region of Nigeria', *Transactions and Papers of the Institute of British Geographers*, vol. xxiii (1957), 213–24.

— 'Settlement Patterns of the Eastern Region of Nigeria', *Nigerian Geographical Journal*, vol. i, no. 2 (Dec. 1957), 23–30.

Ogbaugu, B., 'Enugu – Coal Town', *Nigeria Magazine*, no. 70 (Sept. 1961), 241–51.

Prothero, R. M., 'The Population of Eastern Nigeria', *Scottish Geographical Magazine*, vol. lxxi (1955), 165–70.

— 'Problems of Population Mapping in an Under-Developed Territory (Northern Nigeria)', *Nigerian Geographical Journal*, vol. iii, no. 1 (Dec. 1959), 1–7.

Udo, R. K., 'Land and Population in Otoro District', *Nigerian Geographical Journal*, vol. iv, no. 1 (Aug. 1961), 3–13.

— 'Patterns of Population Distribution and Settlement in Eastern Nigeria', *Nigerian Geographical Journal*, vol. vi, no. 2 (Dec. 1963), 73–88.

— 'Distribution of Nucleated Settlement in Eastern Nigeria', *Geographical Review*, vol. lv, no. 1 (1965), 53–67.

VI. GEOLOGY

Books, Monographs, Pamphlets, etc.

Hazell, J. R. T., 'Ground Water in Eastern Nigeria', *Geological Survey of Nigeria*, Report No. 5198 (Enugu, 1961). Cyclostyled.

Mironenko, P. A., *Final Report on the Rural Water Supply Problem in the Cross River Plain, Eastern Nigeria*, Ministry of Works (Enugu, 1962).

Mitchell-Thomé, R. C., *Average Annual Natural Recharge Estimates of Groundwater in Eastern Nigeria, with Comments on Consumption*, Geology Department, University of Nigeria (Nsukka, 1963). Cyclostyled.

Monkhouse, R. A., *The Geology of Eastern Nigeria*, Geology Department, University of Nigeria (Nsukka, 1966). Cyclostyled.

Nigeria, Federal Ministry of Mines and Power, Geologic Division, *An Illustrated Review of the Geologic Survey Compiled from Talks Prepared for the Nigerian Broadcasting Corporation* (Lagos, 1963).

— Geological Survey, *Mining and Mineral Resources in Nigeria* (Lagos, 1957).

— *Minerals and Industry in Nigeria* (Lagos, 1957).

Reyment, R. A., *The Future of Geology in Nigeria* (Ibadan, 1964).

— *Aspects of Geology in Nigeria* (Ibadan, 1965).

Simpson, A., *The Nigerian Coalfield*, The Geology of Parts of Onitsha, Owerri and Benue Provinces. Bulletin No. 24, Geological Survey of Nigeria (Lagos, 1954).

de Swardt, A. M. J., and Casey, O. P., *The Coal Resources of Nigeria*, Bulletin No. 28, Geological Survey of Nigeria (Lagos, 1963).

Periodicals

Anon, 'Water Resources of Nigeria', *Nigerian Trade Journal*, vol. xiii, no. 1 (Jan./Mar. 1965), 2–7.

Hazell, J. R., 'The Enugu Ironstone, Udi Division, Onitsha Province' *Rec. Geological Survey, Nigeria* (*1955*) (Kaduna, 1958), 44–58.

Jacobson, R. R. E., Snelling, N. J., and Truswell, J. F., 'Age Determinations in the Geology of Nigeria with special reference to Older and Younger Granites', *Overseas Geol. Min. Resources*, vol. ix, no. 2 (1963), 168–82.

Ledger, D. C., 'Aspects of Nigerian Hydrology', *Nigerian Geographical Journal*, vol. iii, no. 1 (Dec. 1959), 18–27.

Orajaka, S., 'Geology of the Obudu Area, Ogoja Province, Eastern Nigeria', *Nat. Canad.*, vol. xci, no. 3 (1964), 73–100.

VII. LANDFORMS AND DRAINAGE SYSTEMS

Books

Ainslie, J. R., *The Physiography of Southern Nigeria, and its Effect on the Forest Flora of the Country* (Oxford, 1926).

Periodicals

Allen, J. R., 'Coastal Geomorphology of Eastern Nigeria', *Geologie en Mijnbouw*, 44 Jaarg. Nr. 1 (Jan. 1965), 1–21.

Carter, J., 'Erosion and Sedimentation from Aerial Photographs; a microstudy from Nigeria', *Journal of Tropical Geography*, vol. ii (Apr. 1958), 100–6.

Crowder, M., 'Nigeria's Great Rivers; a Survey of the Great Inland Water Highways – the Niger and Benue', *Nigeria Magazine*, no. 64 (Mar. 1960), 28–55.

Iloeje, N. P., 'The Structure and Relief of the Nsukka–Okigwi Cuesta', *Nigerian Geographical Journal*, vol. iv, no. 1 (July 1961), 21–39.

Iloeje, N. P., 'Geomorphology of Nsukka–Okigwi Cuesta', *Nigerian Scientist*, vol. i, no. 1 (Nov. 1961), 13–23.

Jungerius, P. D., 'The Environmental Background of Land Use in Nigeria', *Tijdschrift van het Koninklijk Nederlandsch Aardrijkskundig Genootschap* Deel 81, no. 4 (1964), 415–37.

— 'The Upper Coal Measures Cuesta in Eastern Nigeria', *Annals of Geomorphology* v (1964), 167–76.

Ofomata, G. E. K., 'Some Observations on Relief and Erosion in Eastern Nigeria', *Revue de géomorphologie dynamique*, 17. année, no. 1. Jan.–fév. 1967, 21–9.

— 'Landforms on the Nsukka Plateau of Eastern Nigeria', *Nigerian Geographical Journal*, vol. 10, no. 1 (June 1967), 3–9.

Pugh, J. C., 'The Volcanoes of Nigeria', *Nigerian Geographical Journal*, vol. ii, no. 1 (June 1958), 26–36.

— 'River Captures in Nigeria', *Nigerian Geographical Journal*, vol. iv, no. 2 (Dec. 1961), 41–8.

— and King, L., 'Outline of the Geomorphology of Nigeria', *South African Geographical Journal*, vol. xxiv (1952).

Thomas, M. F., 'On the Approach to Landform Studies in Nigeria', *Nigerian Geographical Journal*, vol. v (1952), 87–101.

VIII. SOILS

Books, Monographs, Pamphlets, etc.

Grove, A. T., *Land Use and Soil Conservation in Parts of Onitsha and Owerri Provinces*, Bulletin No. 21, Geological Survey of Nigeria (1951).

— 'Soil Erosion in Nigeria', in R. W. Steel and C. A. Fisher, *Geographical Essays on British Tropical Lands* (London, 1956), 79–111.

Inyang, P., *Soil Erosion and Soil Conservation in Eastern Nigeria*, Geography Department, University of Nigeria (Nsukka, 1962). Cyclostyled.

Jungerius, P. D., 'The Soils of Eastern Nigeria', *Publicaties van het Fysisch-Geografisch Laboratorium van de Universiteit van Amsterdam*, no. 4 (1964), 185–98.

Obihara, C. H., Bawden, M. G., and Jungerius, P. D., *The Anambra-Do Rivers Area*, Soil Survey Memoir no. 1, Ministry of Agriculture (Enugu, 1964).

Vine, H., *Nigerian Soils in Relation to Parent Material*, Commonwealth Bureau of Soil Science, Technical Communication, no. 46 (Harpenden, 1949).

— *Notes on the Main Types of Nigerian Soils*, Nigeria Agricultural Department, Special Bulletin No. 5 (Lagos, 1953).

Periodicals

D'Hoore, J., 'La Carte des sols d'Afrique au sud du Sahara', *Pedologie*, x (1960).

Floyd, B. N., 'Soil Erosion and Deterioration in Eastern Nigeria: A Geographical Appraisal', *Nigerian Geographical Journal*, vol. viii, no. 1 (June 1965), 33–44.

Grove, A. T., 'Farming Systems and Soil Erosion on Sandy Soils in South Eastern Nigeria', *Bulletin Agric. du Congo Belge*, xl, 2150–5.

— 'Soil Erosion and Population Problems in South-East Nigeria', *Geographical Journal*, vol. cxvii (1951), 291–306.

Jungerius, P. D., and Levelt, T. W. M., 'Clay Mineralogy of Soils over Sedimentary Rocks in Eastern Nigeria', *Soil Science*, vol. xcvii, no. 2 (Feb. 1964), 89–95.

Obihara, C. H., 'The Acid Sands of Eastern Nigeria', *Nigerian Scientist*, vol. i, no. 1 (1961), 57–64.

Ofomata, G. E. K., 'Soil Erosion in the Enugu Region of Nigeria', *African Soils*, ix, no. 2 (1964), 289–348.

— 'Factors of Soil Erosion in the Enugu Area of Nigeria', *Nigerian Geographical Journal*, vol. viii, no. 1 (June 1965), 45–59.

Randall, P. C., 'Soil Degradation and Land Use in Onitsha Province', *Farm and Forest*, vol. i, no. 2 (1940), 21–5.

Sykes, R. A., 'A History of the Anti-Erosion Work at Udi', *Farm and Forest*, vol. i, no. 1 (1940), 3–6.

Vine, H., 'Experiments on the Maintenance of Soil Fertility at Ibadan, Nigeria', *Journal of Experimental Agriculture*, vol. xxi (1953), 65–85.

IX. CLIMATE

Books, Monographs, Pamphlets, etc.

Garnier, B. J., *Using Potential Evapotranspiration as a Guide to the Water Requirements of Crops*, University of Ibadan (Ibadan, 1957).

— *Weather Conditions in Nigeria*, Climatological Research Series No. 2 (McGill University, Montreal, 1967).

Gregory, S., *Rainfall over Sierra Leone*, Department of Geography, University of Liverpool, Research Paper No. 2 (Liverpool, 1965). Section 1: West African climate.

U.K., British West African Meteorological Services, *Preliminary Note on the Climate of Nigeria* (1950). Cyclostyled.

— *Gusts in Nigeria*. Nigerian Meteorological Notes No. 1 (Lagos, 1952). Cyclostyled.

University of Nigeria, Faculty of Agriculture, *Basic Data: Agrometeorological Station* (Nsukka, 1964). Cyclostyled.

Walker, H. O., *The Monsoon in West Africa*, Ghana Meteorological Department, Note No. 9 (Accra, 1958).

Periodicals

Crowe, P. R., 'Wind and Weather in the Equatorial Zone', *Transactions of the Institute of British Geographers*, xvii (1951).

Davies, J., 'Estimation of Insolation for West Africa', *Quarterly Journal, Royal Meteorological Society*, vol. xci, no. 389 (July 1965), 359–63.

— 'Solar Radiation Estimates for Nigeria', *The Nigerian Geographical Journal*, vol. ix, no. 2 (Dec. 1966), 85–100.

Eldridge, R. H., 'A Synoptic Study of West African Disturbance Lines', *Quarterly Journal, Royal Meteorological Society*, vol. lxxxiii, no. 357 (1957), 303–14.

Ene, N., 'A Bibliography on the Climate of Nigeria', *Nigerian Geographical Journal*, vol. v, no. 1 (June 1962), 53–60.

Garnier, B. J., 'Some Comments on Measurements of Potential Evaporation in Nigeria', Department of Geography, University of Ibadan, *Research Notes*, no. 2 (1953), 11–18.

— 'Report on Experiments to Measure Potential Evapotranspiration in Nigeria', Department of Geography, University of Ibadan, *Research Notes*, No. 8 (1956).

— 'Some Comments on Defining the Humid Tropics', Department of Geography, University of Ibadan, *Research Notes*, no. 11 (1958), 9–125.

— 'Maps of the Water Balance in West Africa', *I.F.A.N.* Series A, vol. xxii, no. 3 (1960), 709–22.

— Fosberg, F. R., and Kuchler, A. W., 'Delimitation of the Humid Tropics', *Geographical Review*, vol. li, no. 3 (July 1961).

Hamilton, R. A., and Archibold, J. W., 'Meteorology of Nigeria and Adjacent Territory', *Quarterly Journal, Royal Meteorological Society*, vol. lxxi, no. 309 (1945), 231–66.

Hare, F. K., 'The Concept of Climate', *Geography*, vol. li, no. 2 (Apr. 1966), 99–110.

Hodder, B. W., 'A Note on Delimiting the Humid Tropics: the case of Nigeria in West Africa', Department of Geography, University of Ibadan, *Research Notes*, No. 10 (June 1957), 1–7.

Ireland, A. W., 'The Little Dry Season of Southern Nigeria', *Nigerian Geographical Journal*, vol. v, no. 1 (June 1962), 7–20.

Ladell, W. S. S., 'Physiological Classification of Climates illustrated by reference to Nigeria', *Proceedings*, International West African Congress, Ibadan, 1949 (Jos, 1956), 4–21.

Miller, R., 'The Climate of Nigeria', *Geography*, vol. xxxvii, no. 4 (Nov. 1952), 198–213.

Pugh, J. C., 'Rainfall Reliability in Nigeria', *Proceedings*, 17th International Geographical Congress (Washington, D.C., 1952), 280–5.

Schove, D. J., 'A Further Contribution to the Meteorology of Nigeria', *Quarterly Journal, Royal Meteorological Society*, vol. lxxii, no. 331 (1946), 105–12.

X. VEGETATION

Books, Monographs, Pamphlets, etc.

African Timber and Plywood (Nigeria) Ltd., *A.T.P. Sapele* (Lagos, n.d.).

Aubréville, A. M., 'Tropical Africa', in S. Haden Guest, J. K. Wright, E. M. Teclaff, *A World Geography of Forest Resources* (New York, 1956), chap. 16.

Keay, R. W. J., *An Outline of Nigerian Vegetation* (Lagos, 1959).

Molski, B. A., *The Importance of Shelterbelts as a Means of Developing Nigerian Agriculture*, Botany Department, University of Nigeria (1966). Cyclostyled.

Nigeria, Federal Ministry of Commerce and Industry, *Nigerian Timber* (Lagos, n.d.).

— Federal Ministry of Information, *Some Nigerian Woods* (Lagos, 1962).

— Forest Department, *The Vegetation of Nigeria*; descriptive terms (Lagos, 1948).

Richards, P. W., *The Tropical Rain Forest* (Cambridge, 1952).

Rosevear, D. R., *Check List and Atlas of Nigerian Mammals*; with a foreword on vegetation (Lagos, 1953).

— and Lancaster, P. C., *Our Forests* (Lagos, n.d.).

Periodicals

Adeyoju, S. K., 'The Forest Resources of Nigeria', *Nigerian Journal of Geography*, vol. viii, no. 2 (Dec. 1965), 115–26.

Jones, A. P. D., 'Notes on Terms for Use in Vegetation Description in Southern Nigeria', *Farm and Forest*, vol. vi (1945).

Killick, H. J., 'The Ecological Relationships of Certain Plants in the Forest and Savanna of Central Nigeria', *Journal of Ecology*, vol. xlvii, no. 1 (Mar. 1959), 115–27.

Northern Nigeria, Ministry of Agriculture, 'A Key to Nigerian Grasses', *Samaru Research Bulletin*, no. 1 (Kaduna, 1960).

Rains, A. B., 'A Field Key to the Commoner Genera of Nigerian Grasses', *Nigerian Field*, vol. xxii, no. 4.

Richards, P. W., 'Ecological Studies on the Rain Forest of Southern Nigeria', *Journal of Ecology*, vol. xxvii (1939), 1–61.

— 'The Floristic Composition of Primary Tropical Rainforest', *Biological Review*, vol. xx (1945).

Rosevear, D. R., 'Mangrove Swamp', *Farm and Forest*, vol. viii, no. 1(1947).

XI. AGRICULTURE

Books, Monographs, Pamphlets, etc.

Anomba, E. O., 'Rural Economy of Onitsha Southern County', University of Ibadan, Original Essays in Geography (Ibadan, 1962). Manuscript.

Aurlien, O., 'Research Results from Fertilizer Trials'. Paper presented at Conference on Fertilisers, University of Nigeria (Jan. 1964). Cyclostyled.

Bartlett, R., *Rubber in Eastern Nigeria*, United States Agency for International Development. Consultant Report No. 10 (Lagos, 1961). Cyclostyled.

Christiansen, J. E., *et al.*, *Preliminary Survey of the Cross River Drainage Basin of Eastern Nigeria*, United States Agency for International Development (Enugu, 1963). Cyclostyled.

Coppock, J. T., 'Tobacco Growing in Nigeria', *Erdkunde*, vol. xix, no. 4 (1965), 297–306.

Davis, L. L., *Status of Rice Production in Nigeria*, United States Agency for International Development. Consultant Report No. 12 (Lagos, 1962). Cyclostyled.

Eastern Nigeria Development Corporation, *Oils* (Enugu, n.d.).

— *The E.N.D.C. in the First Decade 1955–1964* (Enugu, n.d.).

Eastern Nigeria, Ministry of Agriculture, *Eastern Nigeria Farm Settlement Scheme*, Agricultural Bulletin No. 2 (Enugu, n.d.).

— *Eastern Nigeria Farm Settlement Scheme*. Supplement to Agricultural Bulletin No. 2. Technical Bulletin No. 6 (Enugu, n.d.).

— *Farm Settlement Scheme*, Third Annual Progress Planning Conference, Technical Bulletin No. 4 (Enugu, 1963).

— *Crop Calendar for Eastern Nigeria*, Technical Bulletin No. 8 (Enugu, 1963).

— *Farm Management Studies in Eastern Nigeria 1964–65* (Enugu, 1965).

— Ministry of Economic Planning, *Plantation Survey of Eastern Nigeria 1961–62*. Official Document No. 17 of 1965 (Enugu, 1965).

— Ministry of Information, *Agriculture* ('Progress' Series No. 1, Enugu, 1960).

Ekejiuba, F. I., *Padi Lands of Abakaliki District*, University of Ibadan, Original Essays in Geography (Ibadan, 1962).

Fennell, M. A., *Brief Summary of Cassava Production in Nigeria*, Ministry of Agriculture and Natural Resources, Ibadan (1961). Cyclostyled.

Floyd, B. N., 'Rural Land Use in Nsukka Division', in P. K. Sircar (ed.), *Nsukka Division: A Geographic Appraisal* (Nsukka, 1965), chap. 5, 51–71. Cyclostyled.

— *ENDC Ranch*: Cattle Ranching in Eastern Nigeria (Enugu, 1965).

Halverson, J. H., *The Present Nutritional Situation in Nigeria*, SSOM/ Nigeria Consultant Report No. 4 (Lagos, 1961). Cyclostyled.

Helleiner, G. K., *Peasant Agriculture, Government and Growth in Nigeria*. Economic Growth Center, Yale University (Homewood, Ill., 1966).

Highsmith, R. M., *et al* 'Shifting Cultivation in Southeastern Nigeria', in *Case Studies in World Geography, Occupance and Economy Types* (Englewood Cliffs, New Jersey, 1961), chap. 1, 3–10.

Johnson, G. J., 'Making Practical (Economic) Fertilization Recommendations'. Paper presented at Conference on Fertilisers, University of Nigeria (Jan. 1964). Cyclostyled.

Jugenheimer, R. W., *A Maize Improvement Program for Nigeria*, United States Agency for International Development. Consultant Report No. 17 (Ibadan, 1962). Cyclostyled.

Lynn, C. W., *Agricultural Extension and Advisory Work*, with special reference to the Colonies, Colonial Office Publication No. 241 (London, 1949).

McClung, J. R., and Maxwell, T. A., *Presenting a Plan for an Agricultural Credit Service for Eastern Nigeria*, United States Agency for International Development. Consultant Report No. 17 (Lagos, 1962).

MacFarlane, D. L., and Oworen, M. A., 'Investment in Oil Palm Plantations in Nigeria: A Financial and Economic Appraisal', Economic Development Institute, University of Nigeria (1965). Manuscript.

Mann, W. S., *A Study in the Economics of Fertilizer Use in Eastern Nigeria*, Technical Bulletin No. 5 (Enugu, 1963).

Martin, A., *The Oil Palm Economy of the Ibibio Farmer* (Ibadan, 1956).

Morgan, W. B., 'The Strip Fields of Southern Nigeria', in Stamp, L. D. (ed.), *Report of a Symposium Held at Makerere College, 1955* (London, 1956), 33–7.

Nicholas, I. D., *Land Use and Speciality Crops. Eastern Nigeria*, United States Agency for International Development, Consultant Report No. C-36 (Enugu, 1963).

Nigeria, Federal Office of Statistics, *Agricultural Sample Survey*, Bulletin 5, 1959/60 Eastern Nigeria (Lagos, 1961).

Ntukidem, A., 'Population and Agriculture in Southeast District of Enyong Division, Uyo Province', University of Ibadan, Original Essays in Geography (Ibadan, 1962). Manuscript.

Nugent, J. B., and Taplin, G. B., *Cooperatives in Nigeria*, United States Agency for International Development. Consultant Report No. 6 (Lagos, 1961).

Nugent, J. and P., *Marketing and Distribution of Imported Goods in Nigeria*, United States Agency for International Development. Consultant Report No. 7 (Lagos, 1961).

Ogbonnaya, C. I. N., 'A Geographic Appraisal of the Eastern Nigeria Development Corporation's Plantations', University of Nigeria, Original Essays in Geography (Nsukka, 1966). Manuscript.

Okeke, P. N. (Minister of Agriculture, Eastern Nigeria), *A More Abundant Life* (Enugu, 1961).

Oluwasanmi, H. A., *Agriculture and Nigerian Economic Development* (London and Ibadan, 1966).

— Dema, I. S., *et al.*, *Uboma: A socio-economic and nutritional survey of a rural community in Eastern Nigeria*, No. 6 of the World Land Use Survey (Bude, Geographical Publications, 1966), 116 pp.

Peters, J. R., *Report on Grain Storage*, Eastern Nigeria. United States Agency for International Development, Consultant Report No. 15 (Lagos, 1962).

de Schlippe, P., *Shifting Agriculture in Africa*, The Zande System of Agriculture (London, 1952).

Schultz, T. W., *Transforming Traditional Agriculture* (New Haven, 1964).

Sly, J. M. A., 'Research Results from Oil Palm Fertilizer Trials, WAIFOR', Paper presented at Conference on Fertilizers, University of Nigeria (Jan. 1964). Cyclostyled.

Smock, D. R., *Agricultural Development and Community Plantations in Eastern Nigeria*, Rural Development Project, Ford Foundation (Lagos, 1965). Cyclostyled.

U.K., *Report of Nigerian Livestock Mission, 1949*, Col. No. 266 (London, 1950).
— *Insect Infestation of Stored Food Products in Nigeria*, Col. Research Publication No. 12 (London, 1952).
United Nations, F.A.O., *Agricultural Developments in Nigeria, 1965–80* (Rome, 1966).
Upton, M., 'Recent changes from subsistence to commercial agriculture in southern Nigeria', *The World Land Use Survey*, Occasional Papers no. 7, 1966, 29–40.
Usoroh, E. J., 'The Agricultural Geography of Ikot Ekpene District', University of Ibadan, Original Essays in Geography (Ibadan, 1962). Manuscript.
Uzozie, L., 'Patterns of Crop Combination, Concentration and Intensity of Production in Eastern Nigeria', Geography Department, University of Nigeria (Nsukka, 1965). Manuscript.
Virone, L. E., 'Awakening in the Uboma Forests', *Geographical Magazine*, vol. xl, no. 3 (July 1967), 214–24.

Periodicals

Bridges, A. F. B., 'The Oil Palm Industry in Nigeria', *Farm and Forest*, vol. vii, no. 1 (1946), 54–8.
Buchanan, K. M., 'The Delimitation of Land Use Regions in a Tropical Environment: An Example from the Western Region of Nigeria', *Geography*, vol. xxxviii (1953), 303–7.
— 'Recent Developments in Nigerian Peasant-Farming', *Malayan Journal of Tropical Geography*, vol. ii (1954), 17–34.
Coppock, J. T., 'Agricultural Geography in Nigeria', *Nigerian Geographical Journal*, vol. vii, no. 2 (Dec. 1964), 67–90.
— 'Agricultural Developments in Nigeria', *Journal of Tropical Geography*, 23 (Dec. 1966), 1–18.
Duru, R. C., 'Improvement of the Rural Economy of Eastern Niger Delta', *University Geographer* (University of Ibadan), vol. iii, no. 2 1962), 9–11.
Fennell, M.A., 'Some Aspects of Research as it Relates to Economic Development in Nigeria', *Proceedings of the Agricultural Society in Nigeria*, vol. i (1962).
Floyd, B. N., 'Terrace Agriculture in Eastern Nigeria: the Case of Maku', *Nigerian Geographical Journal*, vol. vii, no. 2 (Dec. 1964), 91–108.

Floyd, B. N., and Adinde, M., 'The Farm Settlements of Eastern Nigeria', *Economic Geography*, vol. xliii, no. 3 (July 1967), 187–230.

Hanson, A. H., 'Nile and Niger: Two Agricultural Projects', *Public Administration*, vol. xxxviii (1960).

Hardcastle, J. E. Y., 'The Development of Rice Production and Research in the Federation of Nigeria', *Tropical Agriculture*, vol. xxxvi, no. 2 (April 1959), 79–95.

Herrman, R., 'Water Supply in Nutrition in the Humid Tropics, with special reference to Eastern Nigeria', *Yearbook of the Association of Pacific Coast Geographers*, vol. 28 (1966), 17–27.

Hill, D. H., 'Beef Production in Southern Nigeria', *Nigerian Grower and Producer*, vol. i, no. 3 (June/July 1962).

— 'Animal Breeding and Improvement in Nigeria', *Outlook on Agriculture*, vol. iv, no. 2 (1964), 80–5.

Jarrett, H. R., 'The Present Setting of the Oil-Palm Industry, with Special Reference to West Africa', *Journal of Tropical Geography*, vol. xi (Apr. 1958), 59–69.

Johnson, J., 'Agricultural Development in Nigeria', *Nigerian Grower and Producer*, vol. i, no. 1 (Feb./Mar. 1962), 5–7.

Jones, G. I., 'Agriculture in Ibo Village Planning', *Farm and Forest*, vol. vi, no. 1 (1945), 9–15.

Jones, W. O., 'Food and Agricultural Economies of Tropical Africa', *Food Research Institute Studies*, vol. ii, no. 1 (Feb. 1961), 3–20.

Kilby, P., 'The Nigerian Palm Oil Industry', *Food Research Institute Studies*, Stanford Univ., vol. 7, no. 2 (1967), 177–203.

Kreinen, M. E., 'The Introduction of Israel's Land Settlement Plan to Nigeria', *Journal of Farm Economics*, vol. xlv, no. 3 (1963), 535–46.

Mabogunje, A. L., 'Rice Cultivation in Southern Nigeria', *Nigerian Geographical Journal*, vol. ii, no. 2 (Mar. 1959), 59–69.

Mackay, J. H., 'Perspective in Land Use Planning', *Farm and Forest*, vol. v, no. 3 (1944).

Morgan, W. B., 'The Nigerian Oil Palm Industry', *Scottish Geographical Journal*, vol. lxx (1955), 174–7.

— 'The Change from Shifting to Fixed Settlement in Southern Nigeria', Department of Geography, University of Ibadan, *Research Notes*, no. 7 (1955).

— 'Agriculture in Southern Nigeria (excluding the Cameroons)' *Economic Geography*, vol. xxxv, no. 2 (Apr. 1959), 138–50.

— 'The Distribution of Food Crop Storage Methods in Nigeria', *Journal of Tropical Geography*, vol. xiii (Dec. 1959), 58–64.

Moss, R. P., 'Land Use Mapping in Tropical Africa', *Nigerian Geographical Journal*, vol. iii, no. 1 (Dec. 1959), 8–17.

— 'Soils, Slopes and Land Use in a part of South-western Nigeria; some implications for the planning of agricultural development in inter-tropical Africa', *Transactions of the Institute of British Geographers*, vol. xxxii (1963), 143–68.

Okiy, G. E. O., 'Indigenous Nigerian Food Plants', *Journal of the West African Science Association*, vol. vi, no. 2 (Aug. 1960), 117–21.

Oluwasanmi, H.A., 'Agriculture in a Developing Economy', *Journal of Agricultural Economics*, vol. xiv (1960–1).

— and Alao, J. A., 'The Role of Credit in the Transformation of Traditional Agriculture: the Nigerian Experience', *Nigerian Journal of Social and Economic Studies*, vol. vii (1965).

Oyolu, C., 'Is Nigeria a True Agricultural Country?' *Nigerian Scientist*, vol. i, no. 1 (1961), 38–47.

Prothero, R. M., 'Some Problems of Land Use Survey in Nigeria', *Economic Geography*, vol. xxx (Jan. 1954), 60–9.

— 'Recent Developments in Nigerian Export Crop Production', *Geography*, vol. xl (1955), 18–27.

— 'The Sample Census of Agriculture, 1950–51', *Geographical Journal*, vol. cxxi (June 1955), 197–206.

Stamp, L. D., 'Land Utilization and Soil Erosion in Nigeria', *Geographical Review*, vol. xxviii (1938), 32–45.

Udo, R. K., 'The Migrant Tenant Farmer of Eastern Nigeria', *Africa*, vol. xxxiv, no. 4 (Oct. 1964), 326–39.

— 'Problems of Developing the Cross River District of Eastern Nigeria', *Journal of Tropical Geography*, vol. xx (June 1965), 65–72.

— 'Sixty Years of Plantation Agriculture in Southern Nigeria, 1902–1962', *Economic Geography*, vol. iv, no. 4 (Oct. 1965), 356–68.

Welsch, D. E., 'Rice Marketing in Eastern Nigeria', *Food Research Institute Studies*, Stanford Univ., vol. 6, no. 3 (1966), 329–52.

White, H. B., 'The Movement of Export Crops in Nigeria', *Tijdschrift voor Economische en Sociale Geografie*, vol. xv (1963), 248–53.

XII. FISHERIES

Books, Monographs, Pamphlets, etc.

Abaribe, J. E., 'Fisheries in Eastern Nigeria', University of Nigeria, Original Essays in Geography (Nsukka, 1966). Manuscript.

Niven, D. R., *Fisheries Development in the Eastern Region of Nigeria* (Aba, n.d.).

United States, International Co-operation Administration, *Fisheries Survey of Nigeria*. A Summary Report, Consultant Report Series No. 3 (Washington, D.C., 1961).

Periodicals

Longhurst, A., 'A Review of the Oceanography of the Gulf of Guinea', *Bulletin IFAN*, Series A, vol. xxiv, no. 3 (July 1962), 633–63.

Nzekwu, O., 'Banda. The Secret of Ibo Concentration in Maiduguri', *Nigeria Magazine*, vol. lxxix (1963), 248–53.

Zwilling, K. K., 'Farming Fish', *Nigeria Magazine*, vol. xliv (1954), 315–28.

XIII. INDUSTRIES

Books, Monographs, Pamphlets, etc.

Confederation of British Industries, *Nigeria: An Industrial Reconnaissance* (London, 1961).

Eastern Nigeria, Ministry of Economic Planning, *Eastern Nigeria Industrial Directory*, Official Document No. 29 of 1963 (Enugu, 1963).

— *Industrial Enquiry 1961–62*, Official Document No. 6 of 1964 (Enugu, 1964).

— Ministry of Information, *Development* (Industrial) (Enugu, 1960).

Johnson, J. H. E., *Industrial Location in Relation to Nigerian Economic Development*, Geography Department, University of Nigeria (1963). Cyclostyled.

Muojindu, C. O., *Eastern Nigeria Industrial Mapping*, Economic Development Institute, University of Nigeria (Enugu, 1965). Cyclostyled.

Okigbo, P. N. C., *Nigerian National Accounts 1950–57* (Enugu, 1962).

Okpara, M. I., *The Purpose of Industrialization* (Enugu, 1963).

— *The Search for Investment* (Enugu, 1963).

Shell–BP Petroleum Development Co. of Nigeria Ltd, *100 Questions and Answers on the Oil Search* (Owerri, 1957).

— *The Story of Oil in Nigeria*, Independence 1960 (Lagos, 1960).

United States Agency for International Development, *Development of Small Industries in Eastern Nigeria*, The Kilby Report (Enugu, 1963).

Wells, F. A., and Warmington, W. A., *Studies in Industrialization*, Nigeria and the Cameroons (London, 1962).

Periodicals

Anon, 'Nigercem. Nigeria's New Cement Factory at Nkalagu', *Nigerian Trade Journal*, vol. vi, no. 2 (Apr./June 1958).

— 'Industrial Development', *Nigerian Trade Journal*, Special Independence Issue (Sept. 1960), 14–17.

— 'Nigeria's Oil Refinery', *Nigerian Trade Journal*, vol. xi, no. 4 (Oct./Dec. 1963), 152–4.

— 'Rural Industrialization – Timber', *Eastern Nigeria*, vol. i, no. 3 (Dec. 1963), 22–6.

— 'Breweries in Nigeria', *Nigerian Trade Journal*, vol. xii, no. 1 (Jan./Mar. 1964), 10–15.

— 'Tyre Industry in Nigeria', *Nigerian Trade Journal*, vol. xii, no. 2 (Apr./ June 1964), 48–51.

— 'The Nigerian Petroleum Refinery Project', *Trade and Industrial Bulletin*, vol. iv, no. 40 (Nov. 1965), 1.

Anumudu, T. A., 'Modernizing Our Local Crafts', *Trade and Industrial Bulletin*, vol. v, no. 42 (Jan. 1966), 3.

Nwogu, E. D., 'Oil in Nigeria', *Nigerian Geographical Journal*, vol. iii, no. 2 (Nov. 1960), 15–25.

Okigbo, L., 'Sawmill Industry in Nigeria', *Nigerian Trade Journal*, vol. xiii, no. 4 (Oct./Dec. 1965), 154–8.

XIV. TRANSPORT AND COMMUNICATIONS

Books, Monographs, Pamphlets, etc.

Louis Berger Inc., *Calabar–Ikom Highway Project: Federation of Nigeria – Eastern Nigeria* (Harrisburg, Pa., 1962).

Duru, R. C., 'Marketing and Transport in Orlu/Oguta District', University of Ibadan, Original Essays in Geography (Ibadan, 1962). Manuscript.

Hawkins, E. K., *Road Transport in Nigeria*: a Study of African Enterprise (London, 1958).

Hogg, V. W., and Roelandts, C. M., *Nigerian Motor Vehicle Traffic: An Economic Forecast* (London, 1962).

NEDECO (Netherlands Engineering Consultants), *Report on Niger Delta Development* (The Hague, n.d.).

— *River Studies and Recommendations on Improvement of Niger and Benue* (Amsterdam, 1959).

Nigeria, Federal Ministry of Transport and Aviation, *Consideration of the Report by the Stanford Research Institute on Transport Co-ordination in Nigeria* (Lagos, 1962).

— National Economic Council, *White Paper on Transportation* (Lagos, 1964)

Stanford Research Institute, *The Economic Co-ordination of Transport Development in Nigeria* (California, 1961).

United Nations, Economic Commission for Africa, *Transport Problems in Relation to Economic Development in Africa* (Addis Ababa, 1960).

Walker, G., *Traffic and Transport in Nigeria*, The Example of an Underdeveloped Tropical Territory. Col. Research Studies No. 27 (London, 1959).

Periodicals

Anon, 'Civil Aviation', *Nigerian Trade Journal*, Special Independence Issue (Sept. 1960), 28–30.

— 'Development of Inland Waterways', *Nigerian Trade Journal*, Special Independence Issue (Sept. 1960), 49–51.

— 'Railway Development', *Nigerian Trade Journal*, Special Independence Issue (Sept. 1960), 54–5.

— 'Road Construction', *Nigerian Trade Journal*, Special Independence Issue (Sept. 1960), 52–3.

— 'Sea Transport', *Nigerian Trade Journal*, Special Independence Issue (Sept. 1960), 45–8.

— 'Progress of Commercial Aviation in Nigeria', *Nigerian Trade Journal*, vol. xii, no. 4 (Oct./Dec. 1964), 150–4.

— 'Towards Better Services', *Eastern Nigeria*, vol. vii (Sept. 1965), 1–4.

— 'The Missing Link', *Eastern Nigeria*, vol. vii (Sept. 1965), 16–19.

Earl, A. K., 'Nigeria's Ports', *Nigeria Magazine*, no. 72 (March 1962), 26–33.

Udo, R. K., and Ogundana, B., 'Factors influencing the fortunes of ports in the Niger Delta', *Scottish Geographical Magazine*, vol. lxxxii, no. 3 (Dec. 1966), 169–83.

XV. ECONOMIC PLANNING

Books, Monographs, Pamphlets, etc.

Barnet and Reef Associates Inc., *Invest in Eastern Nigeria* (New York, n.d.).

— *Investment Opportunities in Eastern Nigeria* (New York, n.d.).

Eastern Nigeria, *Economic Rehabilitation of Eastern Nigeria*: Report of Economic Mission to Europe and North America (Enugu, 1955).
— *Development Programme, 1958–1962* (Enugu, 1959).
— *Revised Development Programme, 1958–62* (Enugu, 1960).
— *Investment Possibilities in the Eastern Region of the Federation of Nigeria*, Official Document No. 11 (Enugu, 1960).
— *Eastern Nigeria Development Plan, 1962–68*, Official Document No. 8 (Enugu, 1962).
Ford Foundation, *Prospects and Policies for Development of the Eastern Region of Nigeria* (Enugu, 1960).
International Bank of Reconstruction and Development (World Bank), *The Economic Development of Nigeria* (Baltimore, 1955).
— *Current Economic Position amd Prospects of Nigeria*, Report No. AF-17a (1964).
Mba, M. K., *The First Three Years*. A Report of the Eastern Nigeria Six-Year Development Plan (Enugu, 1965).
Nigeria, Federal Ministry of Economic Development, *National Development Plan 1962–68* (Lagos, 1962).
— Federal Ministry of Economic Development, *National Development Plan*, Progress Report 1964 (Lagos, 1965).
Petch, G. A., *Economic Development and Modern West Africa* (London, 1961).
United States, Bureau of Foreign Commerce, *Investment in Nigeria*. Basic Information for U.S. Businessmen (Washington, 1957).

Periodicals

Anon, 'National Development Plan 1962–68. Progress Report', *Nigerian Trade Journal*, vol. xiii, no. 4 (Oct./Dec. 1965), 160–6.
— 'New Thinking in Eastern Nigeria: 1', *West Africa*, no. 2534 (25 Dec. 1965), 1457.
— 'New Thinking in Eastern Nigeria: 2', *West Africa*, no. 2535 (1 Jan. 1966), 7–8.
— 'Backing Nigeria's Future', *West Africa*, no. 2542 (19 Feb. 1966), 197–8.
— 'Nigeria on the Move', *West Africa*, no. 2544 (5 March 1966), 265.
Mordi, C. L., 'The Problems and Planning for Industrial and Commercial Development in Nigeria', *Trade and Industrial Bulletin*, vol. v, no. 43 (Feb. 1966), 1, 3.
Schatz, S. P., 'The Influence of Planning on Development: the Nigerian Experience', *Social Research*, vol. xxvii, no. 4 (1960), 451–68.

Index

Place-names and other features in figures are indicated by F after page number. Similarly, place-names and other features in tables are indicated by T after page number. Small 'n' indicates footnote reference. Subjects of photographs have not been included in this index.